Coal-Oil Mixture Technology

Norman W. Lord • Robert P. Ouellette
Oscar G. Farah
METREK Division of the MITRE Corporation

Paul N. Cheremisinoff
New Jersey Institute of Technology

ANN ARBOR SCIENCE
THE BUTTERWORTH GROUP

230 Collingwood, P.O. Box 1425, Ann Arbor, Michigan 48106

Library of Congress Catalog Card Number 81-69551
ISBN 0-250-40495-8

Manufactured in the United States of America

Butterworths, Ltd., Borough Green, Sevenoaks
Kent, TN15 8PH, England

PREFACE

This book presents a review of the applications of coal-oil mixture combustion with a view of their technical feasibility and economy of practice as a substitute for 100% fuel oil use. Liquid fuels using mixtures of fine coal in oil have been known as colloidal fuel, coal-in-oil and, most recently, coal-oil mixtures (COM). This technology has been of interest since the late nineteenth century and tried on a limited scale since 1930, with promising results. Technical advantages are indicated when compared to oil or coal alone. Economics depend on coal-oil price differentials and the readiness of existing oil-fired boilers to withstand corrosion hazards. Strong economic competition from the full coal-burning option in reconversion cases also must be considered. Coal-oil mixtures in the 25–50% range of coal content have the best economic prospects; the lower limit must be exceeded to qualify for a federal tax break and above the upper limit, extensive added technical difficulties must be overcome to handle high-viscosity mixtures of COM. The concept is indeed very sensitive to the economics of oil pricing and should be of interest for some time.

This book is the result of a technical survey undertaken by The MITRE Corporation and sponsored by Electricité de France of Paris to review the status of coal-oil mixture technology, its feasibility, economics and possible future. Electricité de France is gratefully acknowledged for its support in this work and permission to publish the results and conclusions of this study.

Norman W. Lord
Robert P. Ouellette
Oscar G. Farah
Paul N. Cheremisinoff

Lord **Ouellette** **Farah** **Cheremisinoff**

Norman W. Lord is a broad-spectrum physicist with the MITRE Corporation whose research interests include the structure of solids, ocean acoustics and atmosphere-land interactions. He has consulted extensively for government and industries on a wide variety of problems in operations and estimating technological trends.

He received a BEE from Brooklyn Polytechnic Institute, and earned his MA and PhD in physics from Columbia University. Dr. Lord has been a research physicist at Johns Hopkins University Applied Physics Laboratory, Hudson Laboratories of Columbia University and Travelers Research Center in Hartford, Connecticut. He has written more than 40 technical papers covering these areas and his research, and is co-author of *Heat-Pump Technology, Electrotechnology Vol. 4.*

Robert P. Ouellette is Technical Director of the Environment Division of the MITRE Corporation. Dr. Ouellette has been associated with MITRE in varying capacities since 1969, and has been Associate Technical Director since 1974. Earlier, he was with TRW Systems, Hazelton Labs, Inc. and Massachusetts General Hospital. He graduated from the University of Montreal and received his PhD from the University of Ottawa. A member of the American Statistical Association, Biometrics Society, Atomic Industrial Forum and the National Science Foundation Technical Advisory Panel on Hazardous Substances, Dr. Ouellette has published numerous technical papers and books on energy and the environment. He is co-author of the comprehensive *Electrotechnology* survey series published by Ann Arbor Science.

Oscar G. Farah is Manager of International Programs, Energy Resources and the Environment Division at Metrek, a division of the MITRE Corporation. Before joining MITRE he was a research scientist at the Naval Research Laboratories and the Carlisle Barton Laboratory of Johns Hopkins University, and was a member of the technical staff at Bell Telephone Laboratories and the Western Electric Company. Dr. Farah obtained a BS in mechanical engineering from Purdue University and a PhD in electrical engineering from

Johns Hopkins University. He has published a number of papers on world energy demand and supply, and energy transmission and conservation, and is co-author of the Ann Arbor Science book, *Ethylene—Basic Chemicals Feedstock Material.*

Paul N. Cheremisinoff is Associate Professor of Environmental Engineering at the New Jersey Institute of Technology. He is a consulting engineer and has been a consultant on environmental/energy/resources projects for the MITRE Corporation. A recognized authority on pollution control and alternative energy technologies, he is author/editor of many publications, including several Ann Arbor Science handbooks on pollution and energy, such as *Pollution Engineering Practice Handbook, Carbon Adsorption Handbook, Environmental Assessment and Impact Statement Handbook, Industrial and Hazardous Wastes Impoundment,* and *Environmental Impact Data Book.*

CONTENTS

LIST OF FIGURES

LIST OF TABLES

CHAPTER 1

INTRODUCTION: Background of Current Interest

It is now more than a century since a U.S. patent was granted to Smith and Munsell [1] on liquid fuels using mixtures of fine coal particles in oil. Such mixtures have become known as colloidal fuel, coal-in-oil and, lately, coal-oil mixtures (COM). Throughout the intervening years the basic technology has been tried on a limited scale, but tests in the U.S., Britain and Japan since 1930 have been highly visible. Always the results are very promising: technical feasibility is clear, the technical advantages are all on the side of using a coal-oil mixture (COM) compared to either oil or coal alone, and the economic preference over oil is clear when there is a sufficient price margin between oil and coal. Interest in the technology has sparked government-supported programs when oil supplies appeared to be in imminent danger, as during the two world wars, during the early 1930s in Japan, in 1974 and today. However, the economic advantages usually have not been sufficiently convincing outside of these emergencies to sustain the interest of utilities and industry.

In the past, capital costs for equipment to prepare and use the mixture in a retrofitted boiler at the practicable 50-50 levels almost always outweighed the cost advantages for the mixture, even with the coal price per unit of heating value usually one-third to one-half that of oil. Hence, the rate of return is not sufficiently attractive to evoke a favorable decision when there are alternative investments. With coal prices cheap enough, a new steam-generating plant suited for pure coal burning usually is selected; otherwise, a relatively simple and cheap oil-burner plant is favored for its minimal capital requirement.

Prior to October 1973, the above circumstances held without much change because the price of fuels generally varied along with the price of equipment. In the 1950s and 1960s, oil and natural gas became progressively cheaper so that the economic balance was shifted decidedly in favor of oil- and gas-fired plants. Between 1973 and the end of 1978 the price of oil and, later, gas, rose sufficiently to restore the kind of rough balance that existed prior to the second world war. In addition, a sense of wartime-like urgency was created by the generally accepted realization among industrial and utility leaders that the U.S. no longer had a fully dependable petroleum supply. Hence, interest in the use of coal in oil was revived again.

Parallel sentiments have been expressed in Canada, Britain and Japan, all industrialized countries being fully cognizant of the past developments of the necessary technology. The final turn of the screw may well have taken place in 1979 with the loss of appreciable Iranian oil on the world market and another escalation in its price, at first reflected primarily in the spot market, to a level where coal's heating value price now is only one-fourth that of oil. In addition, it it becoming plain that in the near future fuel prices will escalate much faster than capital equipment prices. Balance probably will be restored when extraction oil prices reach the levels anticipated for oil and gas prepared from coal that requires very heavy investment. From that point on fuel prices would roughly parallel equipment prices, and the ratio of raw coal to oil prices may well favor the use of coal-oil mixtures, with economic advantages clearly matching technical advantages.

The Smith and Munsell patent claimed that a coal-oil mixture was cheaper than oil, easier to handle, more combustible than coal and economical to store. However, it was certainly not cheaper than coal and, therefore, not widely adopted in the prevalent coal-using environment of the time. It did become attractive for steamship power in World War I, however, and was tried by the submarine Defense Association, formed in 1917 by American shipping, insurance and oil companies [2].

The U.S. Navy scout vessel, Gem, fitted with watertube boilers, underwent a full-scale test for sea use. The coal-oil slurry consisted of 33% coal powder (passed through a 200-mesh screen) and 1% lime-rosin soap stabilizer in heavy fuel oil [3]. The British Navy, sorely short of oil, never took the chance to utilize the fuel. One drawback was the higher ash content, which lengthened the visible track of fluid waste left on the sea surface in the vessel's wake. After the war, attempts to develop commercial uses of the fuel in the U.S. both on land and sea were unsuccessful. There was some difficulty maintaining the coal in suspension due to rapid settling of relatively coarse coal particles.

Between 1932 and 1935 the Cunard Steamship Company worked with mining and engineering organizations on both land and sea trials of coal-in-oil use [4-6]. Their objective was to reduce oil imports and expand the market for British coal. One of twelve oil-burning boilers in the S.S. Scythia was fired with a 40% slurry. Similar tests were carried out on the

S.S. Berengaria. The only difficulty encountered was that more frequent cleaning of the oil burners was required. In this same era Japan also explored coal-in-oil usc [7] at its National Fuel Laboratory, the Japan National Railway Technical Institute and the Naval Fuel depot. There were also some tests performed in Germany in 1932 of "Fliesskohle," which consisted of 55% powdered coal and 45% tar oil, which burned well and achieved thermal efficiencies of 70–75%. Economic advantages for all these developments still were not sufficiently attractive and the next surge in development awaited the exigencies of the second world war.

In almost parallel developments, teams at the Bureau of Mines [8] and at Kansas State College [9], working with the Pittsburgh and Midway Coal Company, conducted comprehensive tests to explore the main aspects of the preparation and burning processes:

- grinding of coal,
- mixing coal in oil,
- flow of oil-coal mixtures in pipes,
- stability of colloidal fuels in storage,
- combustion of colloidal fuels, and
- economics of manufacture and use.

The conclusions reached by the two groups, largely working independently of each other, were remarkably similar, a fact that lends credence to the details of their findings. Coal grinding efficiency was found to depend on, without being necessarily proportional to, the size and uniformity of the charge of coal, type and size of grinding agent used, and the extent and rate of the grinding process. Mixing operations were found to be carried out most successfully with a propeller-type agitator. It requires the formation of a vortex in oil into which the pulverized coal is fed. For coal weight concentrations up to 40% the colloidal mixture obeys the fundamental law for viscous fluid flow in the laminar range. Hence, at least up to this concentration, it may be transported in pipes without clogging and fed through oil burner nozzles for combustion. As an aid in stabilizing the mixture, Jonnard [9] found spent alkylation acid and aluminum stearate to have some industrial possibilities. Barkeley [8] could not find a chemical aid but did recommend storage temperatures less than 27°C to allow about nine days before settling of the coal particles. Extending the practical concentration range somewhat over that of the earlier work, both the Kansas State College and Bureau of Mines groups found that combustion of the colloidal fuel up to 40%, if the coal powder was passed at 99% through a 230 mesh (61-μ) openings, was better than, or equal to, that of coal or oil alone.

The presumed advantages of coal-oil mixture burning, common in the discussions of both studies, were as follows:

1. It is cheaper than oil in many localities.
2. It handles like oil in pipes and pumps.
3. It has a burning efficiency that equals or exceeds that of oil.
4. It enables better utilization of storage space than does coal.
5. It assists in conservation of petroleum resources.
6. It eliminates danger of spontaneous combustion in stored coal.
7. It is heavier than water so fires are extinguished more easily. In ship sinkings it would not float to the surface.
8. It has an ash and moisture content less than that of coal.
9. It has a heating value per unit volume higher than either coal or oil alone.
10. It better utilizes low-rank coals.
11. It can be burned in present (1946) oil-burning equipment with efficiencies of 80%.

In the early postwar years there was some concern over the adequacy of then-known petroleum reserves [9,10]. In neither case was the urgency great enough or sufficiently long lasting to cause a price rise in oil that would revive a waning interest in using coal. Mining, transporting and pulverizing coal increasingly have been considered only as a last resort. As soon as the petroleum reserve inadequacy was relieved, the slight economic reason for coal-oil slurry use vanished. Between 1946 and 1973 there was no interest in coal-oil slurries because coal production dropped and its price rose relative to oil. Only in steelmaking, with its requirement for coke in the iron ore reduction process, was there some further trial of coal-oil slurry combustion.

Coke is relatively expensive and normally charged at the top of a blast furnace. In addition to the melt heat, it provides carbon to react with oxygen to produce carbon monoxide (CO) and to react with water to produce CO and hydrogen (H_2). Both of these species then undergo a series of reduction reactions with iron ore (Fe_2O_3) to yield iron. Experiments on the injection of a coal-oil slurry to advantageously provide supplemental carbon in open-hearth operations were carried out in 1944 at ARMCO steel [11]. The practice was discontinued because of sedimentation problems. Rudzki et al. [12] relate that in 1953 Shell Oil developed a stable suspension of by-product carbon black in light residual fuel oil in a 1:1 ratio, using a ball mill to crush and pulverize the carbon black in oil. Carbon particle size was 0.5 μ. In 1955–1956, cooperating with Shell, Bethlehem Steel used the mixture over a range 0% to 10% to obtain improved furnace output (15% increment for 10 wt % carbon

Table 1-1. Comparison of Coal-Oil Mixtures with Other Fuels [12]

Fuel	Composition						Calorific Value		Density			Btu Equivalent
	C %	H %	O %	S %	Ash %	C/H Ratio	(Btu/lb)	(Btu/gal) 81° F	(g/cc) 81° F	(lb/gal) 81° F	(lb/gal) 200° F	(gal/gal of No. 6 Oil, 81° F)
No. 6 Fuel Oil	87.5	10.4	—	1.2	—	8.4	18,520	150,000	0.970	8.10	7.78	1.000
Bituminous Coal[a]	86.5	4.8	5.7	0.5	2.5	18.0	15,290	—	1.300	—	—	—
Liquid Carbon												
20% Coal	87.3	9.3	1.1	1.1	0.5	9.4	17,874	152,640	1.023	8.54	8.25	0.983
30% Coal	87.2	8.8	1.7	1.0	0.8	9.9	17,551	153,922	1.050	8.77	8.50	0.975
45% Coal	87.0	7.9	2.6	0.9	1.1	11.0	17,067	157,016	1.100	9.20	8.90	0.955
Blended Pitch	89.5	7.4	1.8	1.3	—	12.1	16,667	160,000	1.119	9.60	8.95	0.938

[a]Volatile matter, 23.5%.

black). This led to extended work [12] on developing a coal-oil suspension for blast furnace use. However, it was not an accepted practice by the mid-1960s. A typical comparison of the constituency and heating values of such coal-oil mixtures with other fuels is shown in Table 1-1. Since that time there has been increasing interest, as evidenced by trial runs, in using coal-oil slurries to replace oil or gas injection at the bottom tuyeres of blast furnaces, which earlier had become a well-established practice [13].

The advantages of injecting coal at the tuyere had been great enough to impel some trial of coal itself. One such arrangement was described by Dietz [14], as illustrated in Figure 1-1. With this arrangement the Weirton Steel Company successfully injected coal at rates up to 15% of the total furnace fuel requirements. The estimated coal-to-coke replacement ratio was on the order of one pound of coke replaced for each pound of coal injected. Most of the difficulties in Weirton's procedure were in handling the finely powdered coal at the injection stage, which were alleviated by using the coal-oil slurry. Ashland Oil and Refining Company had, in fact, developed a 60-40 coal-oil slurry [15], which it sought to market for blast furnace use. The claim was that its use would "back out" 10–30% of the more expensive coke used in conventional furnace operations. There is also a technical advantage in using coal-oil over oil alone because when oil alone is injected, blast furnace temperatures must be raised about 17°C. With coal-oil, blast temperatures can be about the same as those when coke is the sole source of heat. For each pound of slurry, Ashland estimated 1.4 pounds of coke replaced. At the time (1962), Ashland offered four types of systems ranging in cost from $75,000 to $125,000, with payback in coke savings in three to six months. A typical furnace for application would operate at 1500 ton/day with 16 tuyeres. Coal-oil slurry feed would be 220,500 pounds daily. Price levels were $16/ton for coke, $8/ton for coal and 8¢/gal for fuel oil.

However, both the coal and oil constituents were responsible for backing out the expensive coke, and fuel oil was so cheap that its use without coal continued. All fuel injection into blast furnaces is still considered experimental, although frequently reported. Inland Steel's use of fuel oil in furnaces of 15,000 ton/day capacity was reported by Holowaty et al. [16]. A comparison of the use of natural gas with fuel oil at the Sharon Steel Corporation was reported by Walsh [17].

With the sharp rises in fuel oil and, lately, natural gas prices, the use of a coal-oil slurry to save on oil (since coal was at least as effective as oil in saving on coke) became more economically attractive in steel-making, the industry with the most recent experience in coal-oil mixture use.

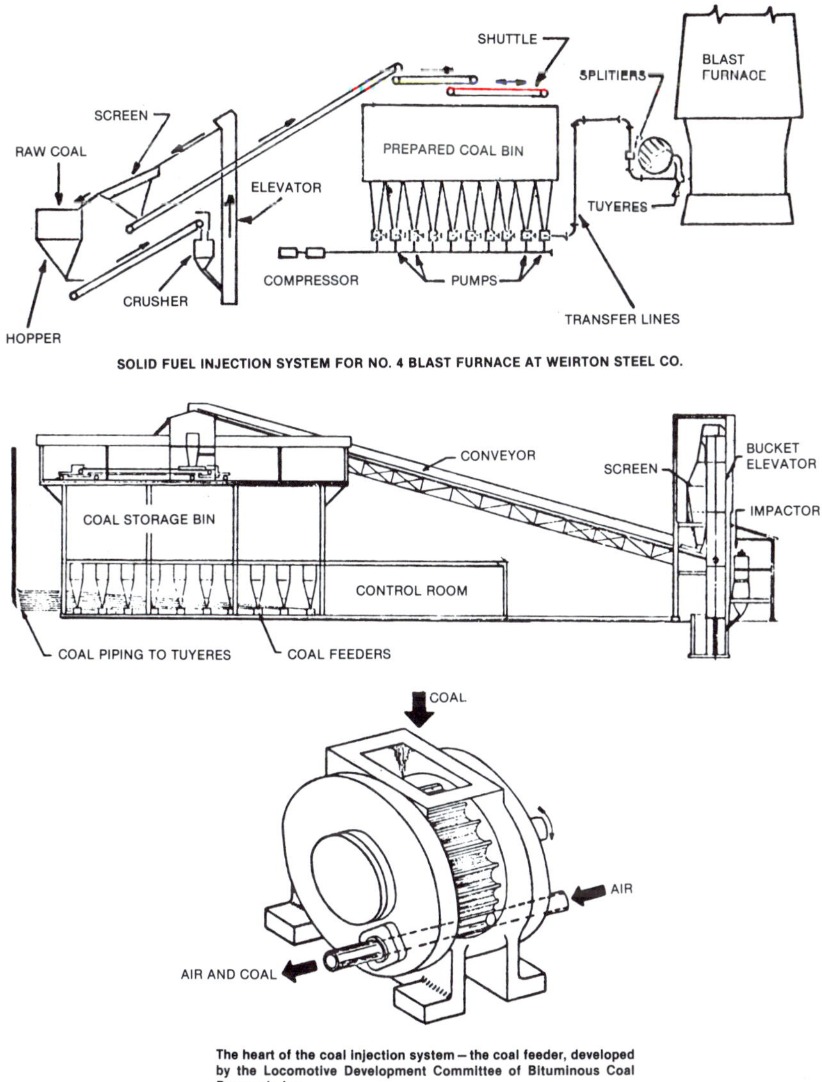

Figure 1-1. Feeding solid coal to blast furnace tuyeres [14].

Hence, in 1976 when the U.S. Department of Energy (DOE) formulated its coal-oil mixtures technology development program [18], its use as auxiliary fuel injection in blast furnaces was included. The project at Interlake has been described by Keyser and Winters [19] as a continuation of Interlake's program, begun in 1973, to reduce injected oil consumption by replacing half of it with coal.

In addition to revived blast furnace use of coal-oil mixtures, General Motors (GM) started testing it for its own industrial heat requirements in 1974 [20]. A pilot test was run for three days on 10,000 gallons of a No. 2 fuel pulverized coal mixture that had been delivered by truck to its plant in Saginaw, Michigan. There were some difficulties with settling of the coal, the mixture's non-Newtonian viscosity characteristics and the cumbersome preparation procedures. By 1975, after some progress in overcoming the difficulties and recognizing that better progress may be had by a broad spectrum of industrial experience, a consortium of cooperating companies was established among fuel suppliers, equipment manufacturers and industrial consumers. The participants, listed in Table 1-2, agreed to make available all information and donate any patents to the public resulting from this work. By 1976 the results were encouraging enough for DOE to expand its program of sponsorship into a number of programs that were aimed to convince practically all operators of industrial and utility gas- or oil-fired boilers to switch to burning a coal-oil mixture.

The main results of the GM project, which was completed in 1977, are as follows:

1. Sufficiently stable COM could be prepared for use in GM's existing industrial boiler originally designed for gas-firing at steam-producing capacity of 120,000 lb/hr. It could be fired up to 75% of rated capacity. The problems remaining in the COM slurry plant were not considered inherent to coal-oil fuels.
2. Air-atomized burners and boiler operated well. Coal carbon burnout was 98%. There was no excessive buildup of ash or slagging on the boiler tubes or on the furnace floor and no sign of boiler tube

Table 1-2. Consortium of Organizations Joining General Motors in Coal-Oil Mixture Technology Exchange, 1974-1977

AMAX Coal Company	New England Power Service
Boston Edison	Peabody Coal
Carbonoyl Company	Petrolite Corporation
Combustion Engineering	PPG Industries
Consolidation Coal	Shell Development Company
DOW Chemical Company	Systems Research Laboratories
Electric Power Research Institute	Texas Utilities
Forney Engineering	Townsend and Bottum
Gulf Research & Development Company	TRW
Island Creek Coal	U.S. Energy Research and Development Administration
Joy Manufacturing	
Kenneth H. White, Inc.	Wayne State University

erosion. The measured stack emission could be treated by standard technology, and boiler efficiency did not suffer.

3. Variations in viscosity of the coil-oil mixture and transient flow constrictions were suspected of causing a fluctuating fuel pressure, which destabilized boiler steam output.
4. Data produced in the study, allowing for installation of necessary emission control devices, indicated that the economics of COM use were marginally attractive in the context of 1977 prices on oil and and coal.

Although the comprehensive investigations have been concluded, General Motors is continuing economic analyses of different sites in New England where GM operates boilers it would like to convert to COM-firing. The DOE has expanded and transferred the GM-led program to the widespread array of new independent projects that provide the primary reference for the current COM state-of-the-art.

Using data primarily supplied by the American Boiler Manufacturers' Association for boiler sales over the 30 years up to 1976, one can plausibly infer accumulated boiler capacities in the gas- and oil-fired categories as indicated in Table 1-3 [18]. Based on this capacity, with load factors of 90% for industrial use and 50% for utility use, considerable savings in oil and gas are potentially available (Table 1-4) if coal-oil mixtures are used for both gas- and oil-fired boilers. A more recent appraisal of prospective U.S. oil savings, based mostly on 50% penetration levels, has been made by Foo [22] at MITRE. He arrives at 373,000 bbl/day, as shown in Table 1-5.

The present DOE program aims to study and develop the best technical approach to the various issues in two major categories: (1) COM preparation and handling, and (2) COM combustion and pollution control. In addition, the economic evaluation of various COM alternatives

Table 1-3. Total Number and Capacity of Industrial and Utility Boilers [21]

	Industrial 100×10^6 Btu/hr or 10 MW	Utility 500×10^6 Btu/hr or 50 MW	Total
OIL—Total Number of Units	1402	457	1859
Total Capacity (Btu/hr)	0.247×10^{12}	0.872×10^{12}	1.12×10^{12}
GAS—Total Number of Units	2905	287	3192
Total Capacity (Btu/hr)	0.691×10^{12}	0.605×10^{12}	1.23×10^{12}
Overall Total—Number of	4307	744	
Units Capacity (Btu/hr)	0.868×10^{12}	1.478×10^{12}	

Table 1-4. Oil Saved by Switching to Coal-Oil Mixture (50 wt %)—Industrial Boilers 100×10^6 Btu/hr and Utility Boilers 50 MW [18]

Fuel Switched	Percent of Capacity Converted: 50% — Oil saved (bbl/yr) (bbl/day)	50% — Coal required (ton/yr)	80% — Oil saved (bbl/yr) (bbl/day)	80% — Coal required (ton/yr)	100% — Oil saved (bbl/yr) (bbl/day)	100% — Coal required (ton/yr)
Oil—50/50 COM	129×10^6 (353×10^3)	29×10^6	206×10^6 (564×10^3)	46×10^6	257×10^6 (706×10^3)	58×10^6
Gas—30/70 COM[a]	115×10^6 (315×10^3)	26×10^6	183×10^6 (501×10^3)	42×10^6	229×10^6 (627×10^3)	52×10^6

[a] As compared to switching to oil.

Table 1-5. Potential U.S. COM Market [22]

Use	Type of Installation	No. of Units	Capacity (MW)	Penetration (%)	Oil Displaced (10^3 bbl/day)
Utility	Designed for coal[a]	121	17,188	50	59
	Designed for oil	321	67,906	50	233
Industrial	Designed for coal[a]	425	16,300	50	34
	Designed for oil	1,291	19,319	25	20
Blast Furnace	Tuyere oil injection	200	50×10^6 bbl/yr oil equiv.	50	27
					373
			Total oil displaced = 373,000 bbl/day		
			Coal required = 33×10^6 ton/yr		

[a]Currently burning oil/gas.

is being studied to guide the formulation of government policy support. The current results of the program in these three areas are discussed in the next three sections. A final section describes the current status of COM development programs. Probably the most significant feature is the lively private interest in the use of COM. One company, CoaLiquids, has established a central plant for COM preparation and is already marketing it to industrial customers.

CHAPTER 2

PREPARATION OF COAL-OIL MIXTURES

Throughout the earlier period of development, emphasis was placed primarily on preparing the coal-oil mixture so that it would be stored for adequate periods of time at atmospheric temperatures and finally fed to the burner nozzles without losing the characteristics that made it a suitable fuel substitute for oil. Despite the minimal extent of tangible application, as outlined above in Chapter 1, there was intense developmental activity on the array of technical issues as shown by the list of relevant patents during the period of 1912-1935. These were compiled by Manning and Taylor [4,5] and are shown in Table 2-1. Those patents filed in both the U.S. and Great Britain are listed as British. The extent of the early U.S. work underlying coal-oil mixture (COM) use on the U.S.S. Gem is shown by the array of patents granted to Lindon W. Bates, who lead the work under the auspices of the Submarine Defense Association. With the initiation of the General Motors-led consortium in 1975 [23], the DOE program for COM preparation was organized along the technical divisions shown in Table 2-2. They follow roughly the specific tasks of the General Motors project listed below, which were outlined in accord with the sequence of activities involved in preparing and utilizing COM:

- selection of fuel oil and coal varieties and pulverizing methods based on heating value, sulfur content, vicosity, ash content, etc.,
- fuel blending with respect to coal, oil proportions and additives to promote stability,
- fuel characterization—properties under temperature cycling and vibration,
- burner operation in atomizing COM as ejected from the nozzle for combustion,
- combustion tests to optimize combustion properties and suggest performance in a full-scale boiler, and
- fuel-handling properties when pumped in a closed-loop burner under various control systems.

Table 2-1. Table of Patents of Processes for Making Colloidal Fuel [4,5]

Author	Patent	Year	Remarks
	British.		
Calvert, G.	14,656	1912	Alkali and glue stabilizer
Plauson, H.	17,729	1913	Fine grinding
Bates, L. W.	149,306	1921	Pitch, lime-resin grease, etc.
Bates, L. W.	153,591	1922	Fire-proofness owing to d $>$ 1
Bates, L. W.	154,538	1922	Tars from water gas or oil
Bates, L. W.	154,605	1922	Removal of ash
Bates, L. W.	155,209	1922	Special apparatus
Armstrong, J. J. V.	155,212	1922	Colloid mill: soap and rubber stabilizers
Bates, L. W.	159,173	1922	Congelation to avoid settling
Bates, L. W.	160,754	1922	Deashing by froth flotation
Bates, L. W.	161,929	1922	Grinding in oil
Calvert, G.	163,797	1922	Circulation to prevent settling
Bates, L. W.	165,418	1922	Resin or soap stabilizers
Bates, L. W.	165,419	1922	Resin or soap stabilizers
Bates, L. W.	165,420	1922	Use of occluded air
Bates, L. W.	165,421	1922	Use of occluded air
Bates, L. W.	165,422	1922	Removal of ash
Bates, L. W.	165,423	1922	Petroleum pitch, etc.
Bates, L. W.	165,425	1922	Two % lime-resin soap
Calvert, G.	165,524	1922	Circulatory system
Greenstreet, C. J.	179,567	1922	Oleates, stearates, paraffin wax
Leadbeater, J. A.	191,201	1923	Fine grinding in colloid mill
Berl, E.	195,408	1923	Various stabilizers
Electrical Improvements, Ltd., and Merz, C. H.	210,147	1924	Low-temperature coke
Marcesche, E. C.	237,959	1925	Anthracene oil and coal or pitch

Koppers Co. (Assees. of H. J. Rose and W. H. Hill)	268,372	1927	See U.S.P. 1,925.005
I. G. Farbenindustrie A.-G.	270,702	1927	For internal combustion engines
Trent Process Corporation	277,659	1927	Heating to decompose fine coal
Alox Chemical Corporation	286,260	1928	Use of alkali salt to make gel
American Coalinoil Corporation and Fairweather, H. G. C.	292,673	1928	Deash and thicken oil by oxidation
I. G. Farbenindustrie A.-G.	227,798	1929	Use of high-frequency current
Strevens, J. L.	323,773	1930	Cracking of coal-in-oil mixture after passing through colloid mill
American Coalinoil Corporation	332,343	1930	Air-blowing
Greenstreet, C. J.	362,424	1931	Distillation of mixture
Standard-I. G. Co. (Assees. of J. M. Jennings)	365,892	1932	Catalytic hydrogenation of mixture
Einstein Electro-Chemical Products, Ltd., and Hatschek, E.	366,128	1932	Protective colloid process
Cunard Steamship Co., Adam, R. A., Holmes, F. C. V., and Perrins, A. W.	396,432	1932	Cracked oil with high carbon residue
Ultramar Co., Ltd., and Bales, L. T.	402,846	1932	Heating and hydrogenation
Radiochemisches Forschungsinstitut G.m.b.H.	406,467	1934	Various stabilizers
Nesfield, A. C.	406,675	1934	Special mixing apparatus
Strevens, J. I., and Milford, W. B.	407,360	1934	Various stabilizers: heat under pressure
Challacombe, R. H. D.	408,684	1934	Mixing machinery
Roberts, A. A.	408,951	1934	Starch stabilizer: boric acid to promote combustion
Cunard Steamship Co., Adam, R. A., Holmes, F. C. V., and Perins, A. W.	409,422	1934	Mixture of tar and cracked oil
Wyndham, S. L.	410,883	1934	Milling process
Universal Oil Products Co.	413,927	1934	Cracking distillation
Maina, J.	416,024	1934	Mixing air and gas or vapor
Radiochemisches Forschungsinstitut G.m.b.H.	417,352	1934	Stabilizers of vaseline, wax, wool fat, etc.
Bristow, W. A., and Pope, P. C.	422,628	1935	Fine grinding
Strevens, J. L., and Hatfield, K. G. S.	426,732	1935	Rubber stabilizer
Bluemner, E.	427,538	1935	Carbonization
Roberts, A. A.	433,348	1935	Combustion controlling reagents
Bluemner, E.	436,380	1935	Cracking and centrifuging to separate gases and vapors

Table 2-1, continued

Author	Patent	Year	Remarks
Roberts, A. A.	438,351	1935	Use of water soluble boron salts to promote combustion
	United States		
Spelman, M.	776,365	1904	Suspension of coal in molasses and alcohol
Greenstreet, C. L.	1,431,225	1922	Air-blowing
Greenstreet, C. L.	1,432,178	1922	Air-blowing
Greenstreet, C. L.	1,623,241	1922	Air-blowing
Griessbach, R., and Eisele, J.	1,681,335	1928	Use of bases
Kern, I.	1,899,811	1928	Heat to inflate coal, and put through colloid mill
Rose, H. J., and Hill, W. H. (Assrs, to Koppers Co.)	1,925,005	1933	Solution of coal in tar oil
	German		
Siemens-Halske A.-G. (Assees. of H. Gerdien)	415,170	1922	Spraying from container under high pressure
A. Riebeck'sche Montan Werke A.-G.	444,420	1925	Phenols and alkali phenates
A. Riebeck'sche Montan Werke A.-G.	453,465	1928	Nitrogenous base phenates
Maschinenbau A.-G., Balcke	574,381	1933	Use of colloid mill
Radiochemisches Forschungsinstitut G.m.b.H.	743,241	1927	*See* B.P. 417,352
	French		
Wyndham, S. I.	764,298	1934	*See* B.P. 410,833

Table 2-2. Technical Issues Addressed by COM Projects for DOE on COM Preparation and Handling [18]

System Function	Approaches/Issues
Coal Pulverization	Dry grinding Wet-grinding ball mill Wet dispenser Ultrafine grinding
Mixing/Stabilization	Coal concentration Homogenization Additives/no additives Coal/oil types
Storage	Long-range stability Pumpability/remixability Continuous mixing Temperature
Pumps/Valves/Pipes	Pumpability Erosion/corrosion Particle deposition/settling
Instrumentation	Applicability of commercial instrumentation Effects of erosion and corrosion Operational reliability
Transportation	Effect of vibration Temperature Pumpability

An array of separate programs, in part sponsored by DOE, were described for details of current progress at the first International Symposium on Coal-Oil Mixtures in 1978. A project very similar to that of General Motors is being carried out by the Electric Power Development Company of Japan [24] using somewhat more coursely ground coal and necessarily different varieties of both coal and oil feedstock. There have been experiments with the use of water [25] and methanol [26] as substantial constituents of a coal slurry for combustion. The following subsections will describe the current status of COM preparation technology following the DOE program areas, with pumps/valves/pipes and transportation combined under conveyance.

PULVERIZING COAL

A colloidal suspension of coal particulates in oil will superficially have a stability that varies inversely with the size of the particles. The main difficulties in the earlier work were in not having readily available the equipment needed for fine pulverizing. Jonnard [9] pointed out that

Bates' experiments on the Gem used very large coal particles, which led to extensive experiments on a stabilizing agent to be added to the oil, called a lime-rosin "fixateur" [2]. By the 1930s, finely pulverized coal was being used routinely for utility coal-fired boilers. The question then was how much expense in grinding was necessary to secure a satisfactory mixture. Manning and Taylor [4,5] surmised from the results to that date that the utility-grade pulverized coal was sufficiently fine. This coal would pass 80–90% through a 200-mesh IMM sieve. Particle size would then range from 0.0025 inch (64 μ) in diameter downward. U.S. and British mesh sizes are very similar, as shown in Table 2-3 which is included for reference to mesh-specified particle size. Particle size would, in fact, be distributed over a range below the mesh opening in ways that depend on the pulverizing process details.

As related by Manning and Taylor [4,5], in the propulsion boiler tests on the S.S. Scythia it was found that a COM mixture containing 40% pulverized coal, passing a 200-mesh sieve, could be handled throughout the existing oil-fuel equipment with no difficulty. However, a more concentrated suspension made the fuel too viscous for convenience. The tests carried out in Germany in 1932 indicate a possible trade-off between increased coal content and finer pulverizing. A "Fliesskohle," consisting of 55% powdered coal and 45% tar oil, was prepared by the Machinenbau AG Balcke. The coal was somewhat finer than in the British tests, passing a screen of 10,000 meshes per square centimeter, which would be equivalent to a hypothetical mesh of 250–260 on the British or U.S. scales. This COM was stable and sufficiently mobile for transfer without difficulty from barrels to a pressure vessel, and burners. Manning and Taylor [4,5] suggest that the stability may be due to a peptizing action of the tar oil on the coal. The General Motors 1975-1977 program tests confirm

Table 2-3. Standard Screen Sizes

U.S. Standard Screen		
Mesh	Inches	Millimeters
20	0.033	0.84
30	0.023	0.59
40	0.165	0.42
50	0.0117	0.30
60	0.0098	0.25
100	0.0058	0.149
140	0.0041	0.105
200	0.0029	0.074
325	0.0017	0.044

W.S. Tyler Screen		
Mesh	Inches	Millimeters
20	0.033	0.83
28	0.023	0.59
35	0.016	0.42
48	0.0116	0.30
60	0.0097	0.25
100	0.0058	0.45
150	0.0041	0.10
200	0.0029	0.074
325	0.0017	0.043

the enhanced stability bestowed by fine grinding, as shown in their tests on 50% coal-oil mixtures [20]. In effect, it was important to assure at least 80% passage through 200-mesh (61 μ) to realize, due to adding homogenized water solutions, a reduction of settling of special additives such as Drew Chemical Corporation's D-5200.

As part of the DOE program, Luckie and Fletcher of the Kennedy Van Saun Corporation investigated the grinding of coal in oil for its inherent advantages over dry grinding [27]. They considered that the grinding process, in which particles would be successively broken down to smaller size by the impacts of falling steel balls, could be represented as a Markov process. That is, the further development of the distribution over particle size after time, t, is completely determined by the distribution at time, t, and the probability function describing the dynamics of particle breakage. The history of particle size distribution changes prior to time is irrelevant. Details on such a representation of grinding as a rate process are discussed by Austin [28]. Luckie attended to the following parametrical features of the concept to gain understanding of the distinctions among various dry and wet coal grinding processes.

We must express the rate at which material of a particular size, y, is selected for redistribution among smaller sizes, x. A selection function

$$S(y)\ dt + 0\ (dt)$$

represents the probability that a new size distribution occurs during the interval $(t, t + dt)$ due to particle breakup by the falling balls. The particle of size y is now broken down to a size x according to the particle progeny function

$$B(x,y)$$

which is a conditional probability distribution for a resultant size x for the particle, provided that before the breakup its size was y.

B(x,y) is a function of present particle sizes and, in addition, primarily the characteristics of the material itself. S(y) is also a function of size but, in addition, a combination of material properties and operating conditions. If it does not vary with time, the grinding is considered to be of first order, i.e., the rate of disappearance of material of a given small interval due to breakage of this size is proportional to the weight of materials present. These functions were analyzed for laboratory tests of different grinding techniques to determine any difference in breakage mechanisms as particle sizes were reduced.

A tumbling ball mill is a rotating tube containing spherical steel balls. It consists of a steel cylinder bolted to heads supported on trunnion bearings at each end, which is rotated about its horizontal axis by a drive. The cylinder is loaded with heat-treated forged steel balls, ¾–4 inches in diameter, which tumble during rotation. Operating rotation of the mill is expressed as a proportion of that critical rate at which the balls would start to centrifuge.

Grinding action will be influenced both by the rotational speed and the amount of coal being ground relative to the volume of the balls. The material is fed in through the feed trunnion and is fractured as it travels along the mill by the stresses imparted to the particles during impact with the tumbling balls. Their length to diameter ratio varies from 0.5 to 3.5. When this ratio exceeds 2 or 2.5, the mills usually are partitioned into two compartments to enhance mass transfer through them.

Operational options of a tumbling ball mill circuit are conventionally limited among such alternatives as wet or dry grinding, according to materially agreed criteria. Dry is taken to mean that surface moisture is less than 3% by weight water, with heated air sometimes used to achieve this condition. Wet means 50% water by volume. According to a rule of thumb, one of the incentives for wet grinding is a reduction in required drive power by about 25% from that of dry grinding for the same product fineness. In terms of the Markov process, representation above the B(x,y) function are expected to be the same, but the wet S(y) function is about 30% higher than that of the dry. In Luckie's experiment the kinetic parameters for 20 × 30 (U.S. sieve) Western Kentucky #9 coal were determined for both dry and wet grinding. The B values were found to be the same and S values in ratio 1.26. Another advantage of wet grinding is that no dust collecting is required. However, the grinding media and liner consumption per ton of product ground is higher.

A study of the kinetic parameters using both batch and continuous tests indicated that the advantages of wet grinding were largely lost by substituting oil for water. Two oils were used, No. 4 and No. 6. The test result S-values and V-values for grinding at 50°C are shown in Table 2-4. In all cases, feedstock was Kentucky #9 coal. The B functions were found to be normalizable according to the functional argument y/x. B-values are largely unaffected as shown in part (b) of the table. Comparing the S values in part (a) shows that grinding in No. 4 fuel oil was 1.31 times faster than dry grinding and 1.12 times slower than grinding in water. Slowest of all grinding is No. 6 fuel oil. Size distributions of the ground product for continuous grinding are shown in Table 2-5 for dry and wet and Table 2-6 for number 4 and 6 fuel oils. Implication of these results is that the rate of breakage of the smaller size particles is much greater

Table 2-4. Kinetic Grinding Parameters [27]

(a) S-values

Environment	S (min^{-1})
Water	1.70
No. 4 Oil	1.52
Dry	1.35
No. 6 Oil	1.15 (1.25 @ 75° C)

(b) B-values

y/x	Environment Dry and Water	No. 4 Fuel Oil	No. 6 Fuel Oil
30	1.00	1.00	1.00
40	0.53	0.50	0.54
50	0.31	0.31	0.37
70	0.20	0.21	0.30
100	0.14	0.16	0.25
140	0.09	0.12	0.21
200	0.06	0.09	0.19
270	0.05	0.07	0.17
400	0.03	0.05	0.14

Table 2-5. Size Distributions for Western Kentucky No. 9 20 × 30 U.S. Mesh Coal for Various Grind Times [27]

Size (U.S. Mesh)	Dry Cumulative Weight Percent Less Than Size 0.20 min	1.00 min	3.00 min	6.00 min
20	100.00	100.00	100.00	100.00
30	23.79	72.43	98.53	99.51
40	13.05	52.54	93.73	99.03
50	7.75	35.70	80.54	98.54
70	5.18	25.47	65.96	92.68
100	3.51	18.10	52.29	81.20
140	2.46	13.01	41.50	66.85
200	1.63	8.96	30.18	55.84
270	1.32	6.73	25.37	42.42
400	0.96	4.91	18.35	30.15

Size (U.S. Mesh)	Wet Cumulative Weight Percent Less Than Size 0.125 min	1.00 min	3.00 min	6.00 min
20	100.00	100.00	100.00	100.00
30	22.25	79.34	99.40	99.95
40	12.05	57.60	95.30	99.90
50	7.40	43.25	89.12	99.20
70	4.85	30.62	78.24	96.84
100	3.30	22.72	65.16	90.66
140	2.35	16.95	47.49	80.09
200	1.52	11.90	36.33	67.11
270	1.17	9.40	28.58	53.28
400	0.80	6.80	20.35	39.94

Table 2-6. Size Distributions for Western Kentucky No. 9 20 × 30 U.S. Mesh Coal for Various Grind Times in Oil [27]

Size (U.S. Mesh)	No. 4 Fuel Oil Cumulative Weight Percent Less Than Size			
	0.20 min	0.75 min	2.00 min	5.00 min
20	100.00	100.00	100.00	100.00
30	30.00	65.00	95.00	99.76
40	15.83	45.18	85.83	99.47
50	10.00	31.79	70.38	98.23
70	7.12	23.55	56.18	93.64
100	5.34	18.03	43.29	83.03
140	4.10	14.04	34.40	69.50
200	3.15	10.83	26.98	55.63
270	2.40	8.33	21.00	42.37
400	1.80	6.20	16.14	32.63

Size (U.S. Mesh)	No. 6 Fuel Oil Cumulative Weight Percent Less Than Size			
	0.30 min 0.275	1.00 min 0.92	3.00 min 2.76	5.00 min @ 50°C 4.60 @ 75°C
20	100.00	100.00	100.00	100.00
30	29.73	69.58	93.00	97.27
40	16.77	49.16	80.00	90.00
50	12.02	37.00	67.80	79.10
70	9.63	29.16	53.93	67.80
100	8.16	24.05	45.00	57.03
140	7.08	20.23	38.63	49.50
200	6.25	17.15	34.00	43.86
270	5.64	15.57	29.26	39.23
400	4.70	13.72	26.00	35.00

in water than in the oils. Kinetic studies are being continued to identify the principal effects to learn how oil results could be improved.

For a pilot plant test, the Electric Power Development Company of Japan has chosen a wet ball mill system [24]. Its evaluation of four types of mills is shown in Table 2-7. The system itself is shown in Figure 2-1, and the grinding results, as shown in Figure 2-2, indicate an adequate fineness with about 70–80% 200-mesh pass. Favorable features of the system are as follows:

1. The mill simultaneously will pulverize and dry the coal and mix it with oil.
2. The activity of the coal surface is maintained at a high level since

Table 2-7. Evaluation of Wet Type Mills for COM Preparation [24]

	Ball Mill	Vibration Ball Mill	Tower Mill	Atliter Mill
Raw Coal Size	20 mm	10 mm	20 mm	0.5 mm
Coal Size in COM	-200^{mesh} 80%	-200^{mesh} 80%	-200^{mesh} 80%	20 μm
Input Power (kwh/t-COM)	10	11	8	130
Potential for Scaling Up	Big machine (200 ton/hr) can be made	Big machine can hardly be made (max 5 t/h)	Big machine has not been developed and no experience of operation	Big machine can hardly be developed
Mixing Characteristics	Good	Good	Good	Good
Heating of Mill	Easy	Rather difficult	Rather difficult	Rather difficult
Control of Coal Content	Easy	Easy	Difficult	Difficult
Treatment of Exhaust Gas	Easy	Not Easy	Not easy	Not easy
Detection of Mill Level	Easy	Easy	Difficult	Difficult
Necessity of Classifier	No	Yes	No	Yes
Necessity of Slurry Pump	No	No	Yes	Yes
Erosion	Less severe	Most severe	Severe	Less severe
Maintenance	Rather easy	Not easy	Not easy	Not easy
Prevention of Noise and Vibration	Possible	Possible	Possible	Possible

the coal is not exposed to air. This allows the full capacity of the surfactant to be utilized.

3. The entire system is expected to use less power than a dry mill.

It is helpful to note also that the early comprehensive work of Jonnard [9] and his co-workers compared experiments on different size balls and other shapes are grinding media. Their qualitative recommendations, which would still pertain to dry grinding, and probably also to wet grinding, are as follows:

1. The size of the medium governs the size of the particles of the product.
2. Greater amounts of a smaller size are produced by using smaller balls when used with a proportionately fine feed.

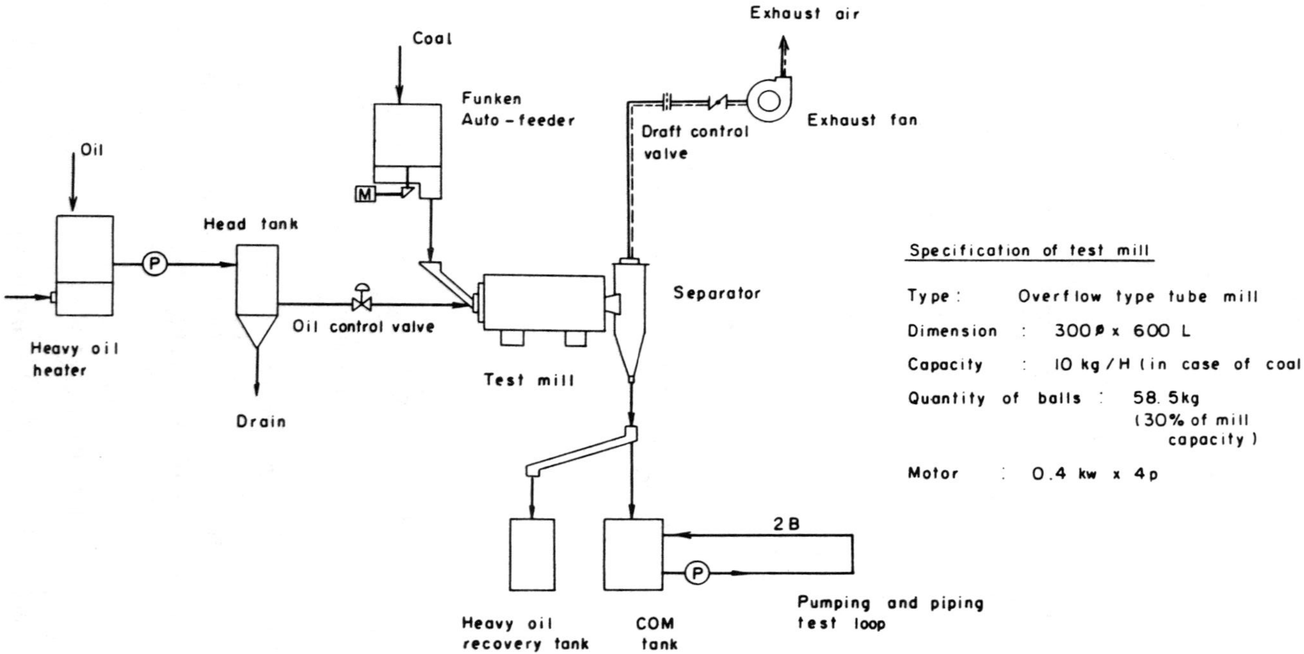

Figure 2-1. Kawasaki Heavy Industries, Ltd. (KHI) wet-type mill test flowchart [24].

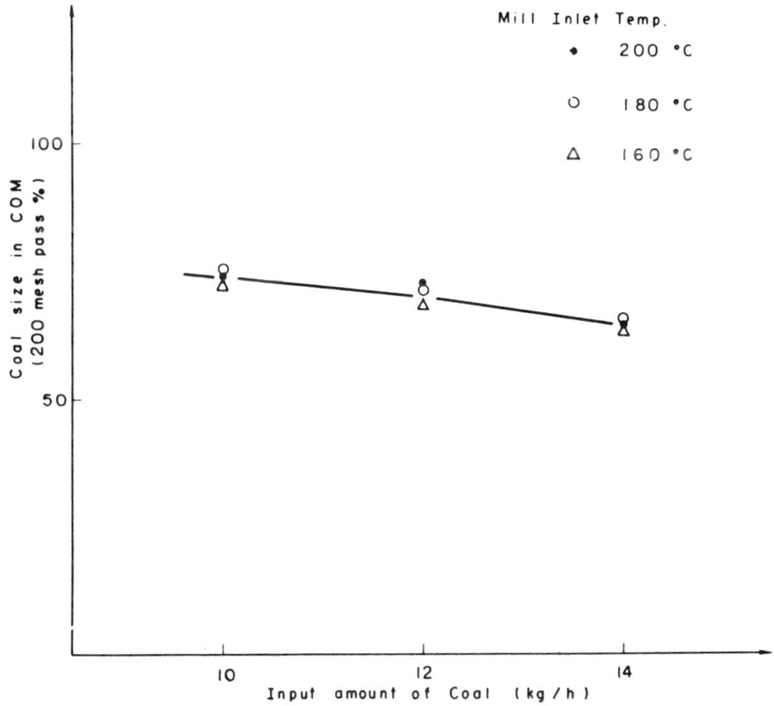

Figure 2-2. Coal ground size according to KHI wet-type mill [24].

3. Any given size of grinding medium operates most efficiently within a certain size range.
4. Mixed media often were a distinct advantage, each size operating most efficiently on a different size of coal.

FORMING THE OIL- OR WATER-COAL COLLOID

Mixing the powdered coal with oil to form a sufficiently stable colloid-like mixture is probably the most critical step in the entire COM utilization procedure. As pointed out by Manning and Taylor [4,5], a suspension of powdered coal in coal usually will be unstable with progressive deposition of coal particles at rates dependent on

- size of the coal particles,
- viscosity of the oil,
- concentration of the suspension, and
- specific gravities of the components.

Stability of the colloid, considered as its resistance to settling out of coal particles, usually can be promoted by some combination of the following tactics:

- finer grinding of the coal,
- establishing a gel structure in the oil with a chemical additive, called "fixateur" or stabilizer,
- using a peptizing agent (electrolyte to provide a double layer of ionic charge around each particle to promote short range ordering with the molecules of the liquid), and
- selecting oil that gives a stable suspension without the need of adding a stabilizer.

A dilute gel will flow like a liquid when subjected to the usual forces involved in handling the oil. However, as originally described by Sheppard [29], it behaves like an elastic solid when subjected to the very small distorting forces that might be exerted by a small descending coal particle. Manning and Taylor [4,5] cite this condition when amounts between 0.2% and 1% of soap or sodium stearate are added to oil. Jonnard [9] cites difficulty in finding "fixateurs," recommending aluminum stearate as best. The Cunard Co. investigations [6] showed that certain oils produced in the cracking process could also form stable suspensions.

Most of the recent work on promoting stabilization has been in the search for additives to create the gel-like structure. A first-order approximation [30] of the required yield strength in the gel shows the direct dependence on particle size as

$$T_y = D_p\, g\, (d_s - d_l) \div (3g_c)$$

where
T_y = required yield strength
D_p = particle diameter
g = gravity acceleration
g_c = gravitational constant
d_s = density of solid particles
d_l = density of liquid carrier

Values of yield strength required to prevent settling of spherical coal particles (density of 1.26 g/cm^3) in a water solution are shown in Figure 2-3. Shape effects introduce extra hindrance, which permits the use of even lower values of yield strength so that those in the figure may be considered maximal.

There are essentially two main ranges of COM coal concentration that are the subjects of current investigations. The first is that based on the classic work centered on a moderate level of 40% coal, at which oil slurries of fine particles seem to achieve a natural stability and above which

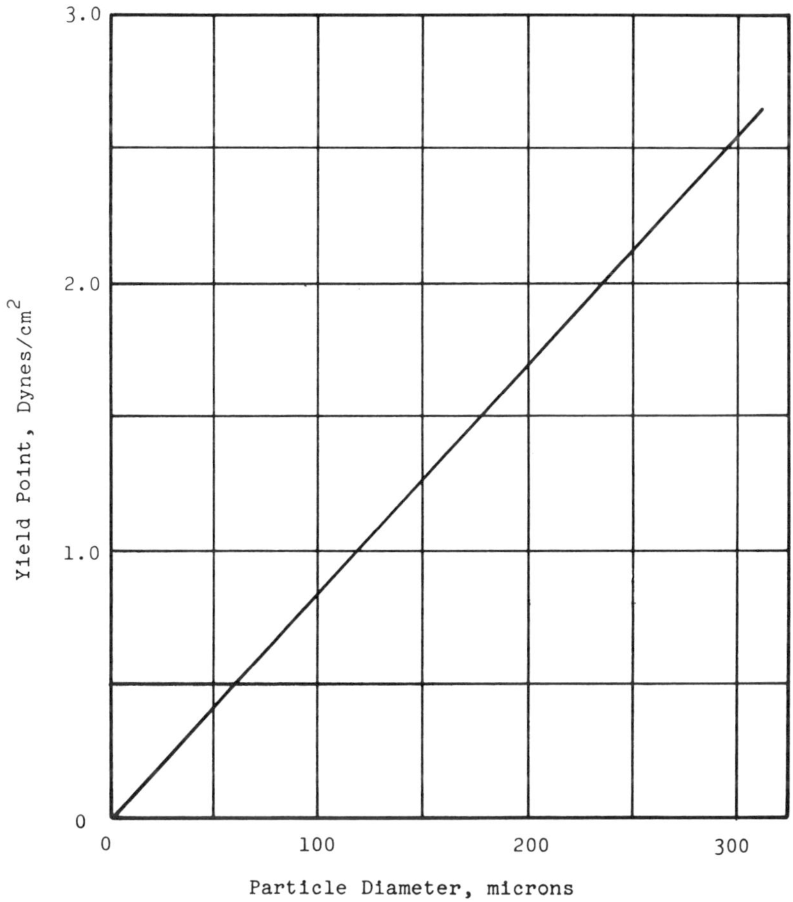

Figure 2-3. Minimum value of yield stress required to prevent unhindered settling of spherical particles [31].

viscosity escalates to an unacceptable level (see the section entitled Viscosity and In-plant Handling). A second, more developmental range includes those high concentrations approaching 70%, at which level oil may be replaced by water and a much larger economic gain may be possible.

COM of Moderate Concentration

The linear dependence of required yield strength on particle size undoubtedly is the underlying reason for Barkeley's [8] finding that stability of the COM depends primarily on fineness of the coal and secondarily on viscosity as a function of temperature. With no fixateur, satisfactory results were obtained with coal particules at 99% passage through a 230-mesh screen (61-μ size) and storage temperatures below 80°F. The Barkeley group tried Bates' lime-rosin grease, 0.1% sulfuric acid (H_2SO_4) and small amounts of coal tar, finding increasing stability in that order. Their results still present one of the most comprehensive portrayals of stability dependence on particle size, temperature and additives. They are shown in Figure 2-4. Stability was maintained for about nine days. Rudzki et al. [12] studied sedimentation rates for much finer particles and their results are shown in Figure 2-5.

A fairly comprehensive recent study of the influence of a number of additives on settling in COM has been carried out by Ekmann and Bienstock [32]. Their test procedure consisted of preparing a COM mixture in a mixing tank with high homogeneity, then transferring it to a settling column housed in a constant-temperature enclosure. Figure 2-6 outlines the physical arrangement. Tests are run in two columns using nonintrusive measurements. In one column, ultrasonic density measuring systems are located 9 and 15 inches from the top. The second column uses gamma radiation density, measuring with one sensor 9 inches from the top of the COM and a lower sensor 5 inches from the bottom. There are thermocouples located at each sensor location and, in each column, four valved sample ports, one at each sensor, one at the bottom and a fourth at an intermediate location. Correlation between the instrument readings and pycnometer samples, drawn at the ports, has been satisfactory. Figure 2-6 outlines the physical arrangement.

Data from the settling tests are presented in Table 2-8. For each test the following two numbers are given:

- $(\Delta SS/\Delta T)_5$—time required for the concentration at the upper sensor to change from 30% suspended solids to 25% suspended solids divided into that 5% change; and
- $(\Delta SS/\Delta T)_f$—percentage change in suspended solids for the test divided by the total time for the test.

Only one of these additives, ICI IL-540, achieved complete stabilization of the mixture. Several others were successful in slowing the settling rate, while others had little apparent, or even a detrimental, effect.

Figure 2-7 presents the temporal change in concentration for a nonstabilized mixture and two mixtures with additives, ICI IL-540 and BASF U-14734. The first of these was prepared by mixing a 0.15% by weight mixture of the additive and 5% water with 65% oil and 30%

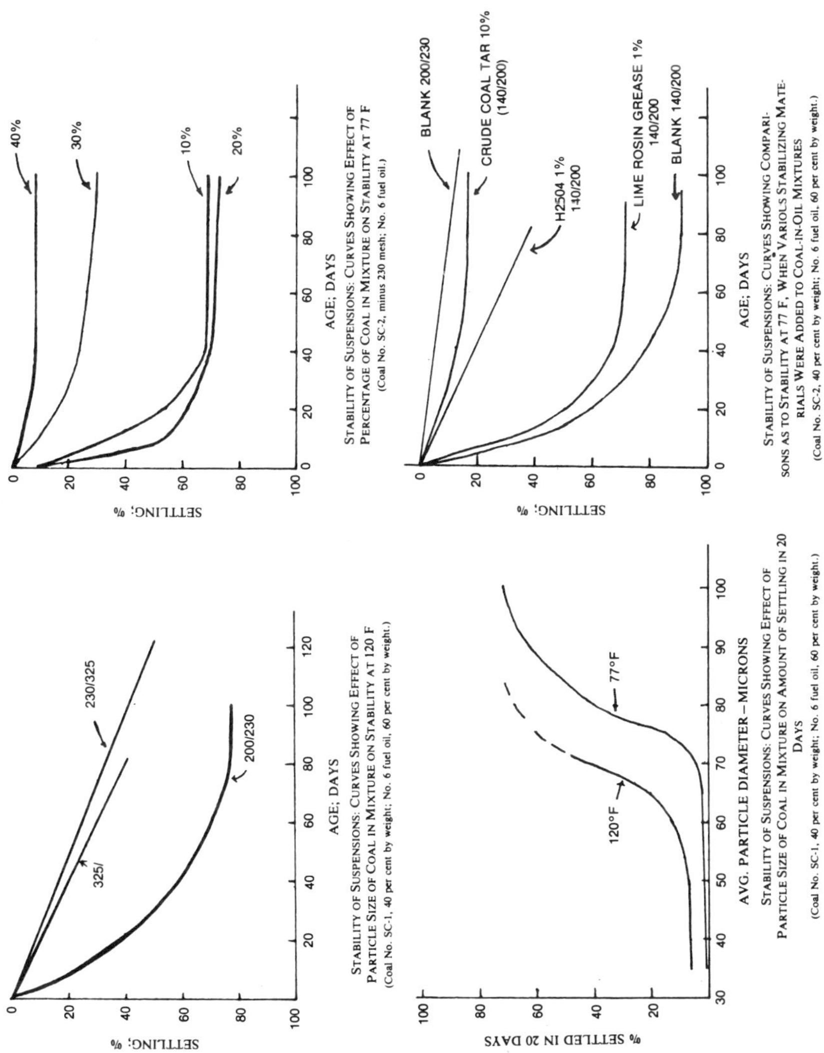

Figure 2-4. Stability of COM suspensions [8].

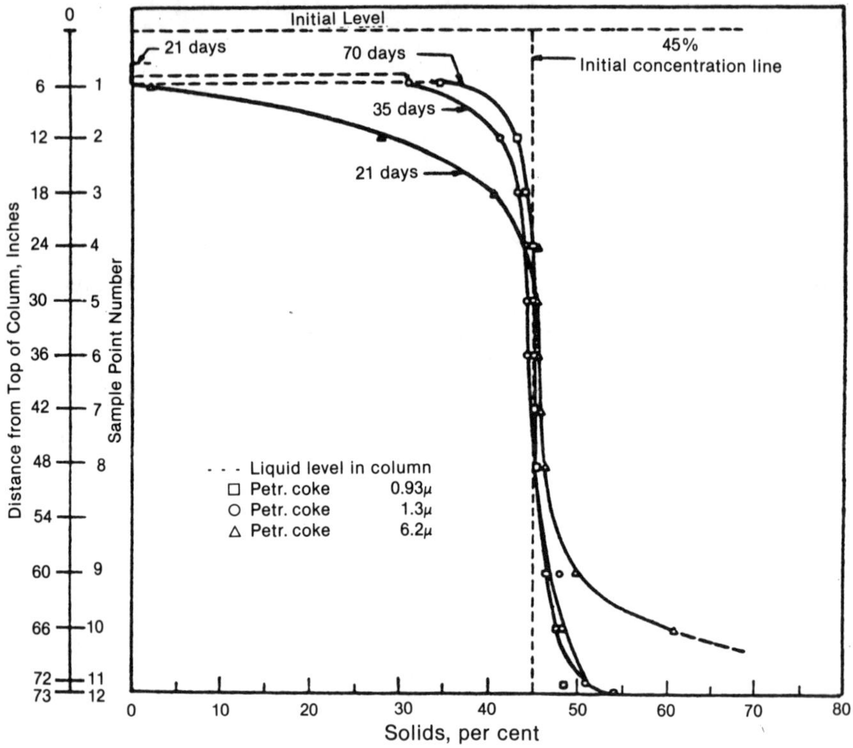

Figure 2-5. Effect of particle size on sedimentation rate at 100°C with 45% solids [12].

coal. The second was prepared by mixing 0.2% of the slurry weight of U-14734 with 3.0% water, 67% oil and 30% coal. Figure 2-8 presents settling data at one sensor location as a function of time for three initial coal concentrations. The 50% COM, without additives, shows no settling after 93 hours, and that mixture remained fluid at all points in the column throughout the test. The 40% COM shows intermediate behavior.

As a comparison with later results on stability, the results of Manning and Taylor [4,5] are noteworthy for their emphasis on the 40% COM. These are shown in Table 2-9, which includes properties of the various oils used. The sodium stearate fixateur appears to confer a stability to the mixture that is midway between that of heavily cracked and normally cracked oils. It was also pointed out that the method dispersing the stabilizing agent is also important. Sodium stearate optimally should be dissolved in oil at 200°C and the mixture allowed to cool. With 0.5% sodium stearate so dispersed, a 40% mixture with Iranian oil showed no separation after standing at atmospheric temperatures for five months.

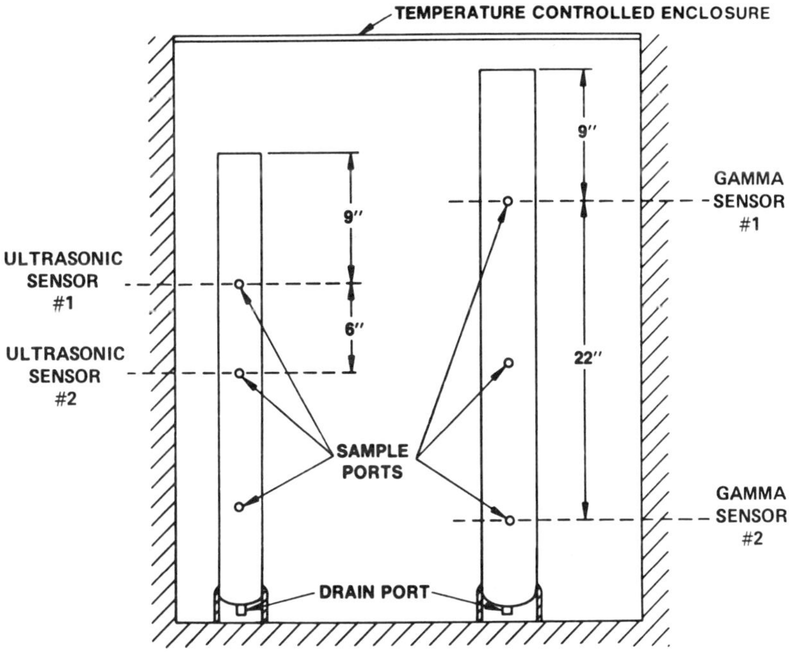

Figure 2-6. Diagram of settling columns [32].

Table 2-8. Effect of Additives on Settling [32]

Test No.	Additive	$(\Delta SS/\Delta T)_5$[a]	$(\Delta SS/\Delta T)_F$[b]	Final % SS	Agglomerated[c]
3	None	0.25	0.12	19	Yes
24	None	—	0.10	17	Yes
19	ICI IL-540	0.00	0.00	30	—
13	BASF 7312	0.04	0.05	24	Yes
7	DS FUELSPERSE 092	0.06	0.06	23	—
23	DS TENLO 70	0.07	0.06	24	Yes
21	EAGLE CRUSHER #1	0.17	0.06	21	Yes
5	DS FUELSPERSE 55	0.06	0.07	24	—
22	EAGLE CRUSHER #2	0.10	0.07	22	Yes
14	BASF U-14734	0.13	0.07	24	Yes
10	KELCO SOLOID	0.33	0.07	22	No
25	DOW XF 43130	0.13	0.08	21	Yes
12	BASF 7071	0.11	0.26	0.0	No

[a]$(\Delta SS/\Delta T)_5 = (5.0\%\ SS)/(\text{time for } 5\%\ SS \text{ drop})$
where SS = suspended solids, T = time.

[b]$(\Delta SS/\Delta T)_F = (30\% - \text{final } \%\ SS)/(\text{total elapsed time for test})$
where SS = suspended solids, T = time
total time between 75 and 125 hours.

[c]Defined as lack of flowability of the COM from the bottom sample ports at the end of the run.

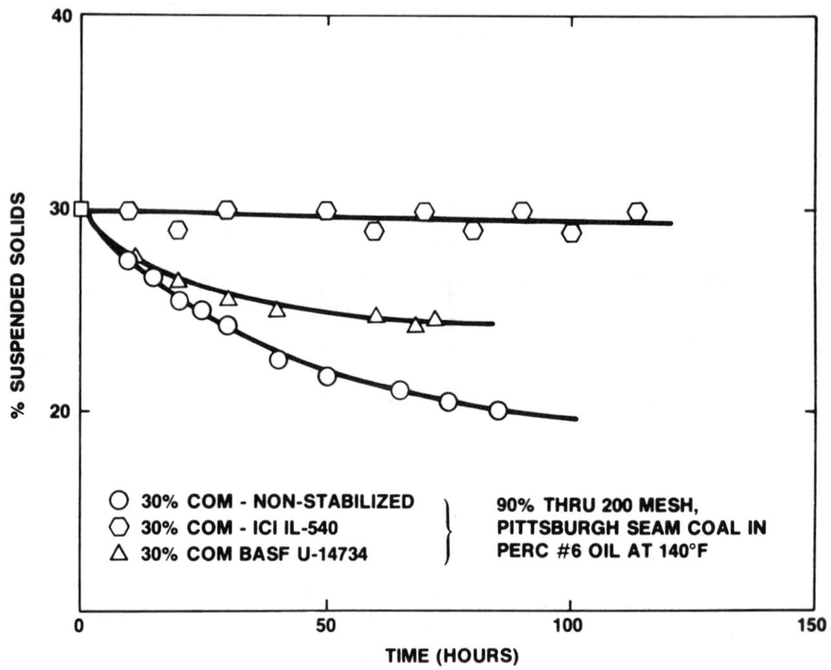

Figure 2-7. Effect of additives on 30% COM [32].

Table 2-9. Stability of COM in Various Oils [4,5]

Description		Oil A Persian Fuel Oil	Oil B Mexican Fuel Oil	Oil C Heavily Cracked	Oil D With 0-25% Sodium Stearate	Oil E Normally Cracked	Oil F High-Petroleum Resin Content
Specific Gravity		—	—	—	0.896	0.925	—
Viscosity (Redwood No. 1)							
at 20° C	secs.	212	17,240	156	693	790	5,500
at 50° C	secs.	62	1,230	63	130	190	480
Carbon Residue %		3.6	9.9	3.8	3.5	3.6	11.0
Hard Asphalt %		0.8	2.4	0.4	1.0	2.1	0.7
Stability of 40: 60 Coal-Oil Mixture—							
At 15-20° C		2-3 weeks		3 weeks	At least 6 weeks	At least 3 months	Permanent (?)
At 50° C		Not tested			5 days	At least 6 days	

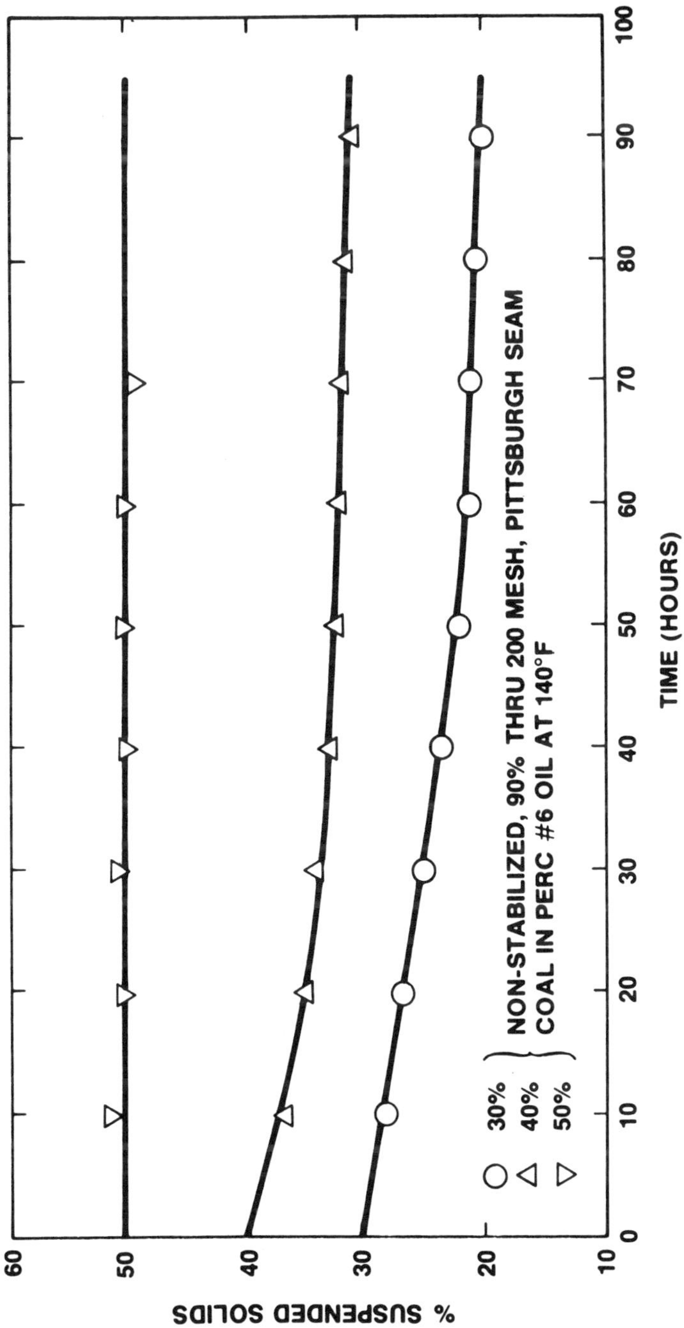

Figure 2-8. Settling behavior as a function of coal concentration [32].

It finally comes down to a question of economics. The soaps and high-temperature dispersal methods are clearly too expensive. The various patents indicate many cheaper additives such as waste rubber, fats and waxes, pitches and bitumen, alkali salts, etc. One simple method appears to consist of blowing air through heated oil to form oxidation products in the oil, which has a stabilizing effects.

COM of High Concentration

An unusual additive, attapulgite clay-surfactant gel, was used by Sawyer [33] of the Pennsylvania Glass Sand Corporation. It was applied to stabilize suspensions of 50–60% coal (200-mesh) in No. 6 fuel oil. The suspension exhibited uniformity when stored for extended periods at 160°F (70°C). The clay is supplied as a powder consisting of agglomerated needles. To obtain the maximum effectiveness from the clay it must be dispersed into its ultimate needles. This is done with a surfactant and medium- to high-shear mixing. Sawyer noted that there appeared to be an energy input "threshold" for the mixing that had to be surpassed to achieve gelling. In his laboratory studies, a Sterling Multimixer (medium shear) and a Waring Blender (high shear) were used. The threshold could be lowered by such factors as oil viscosity, oil type, a functional surfactant, increased temperature and free moisture on the clay.

Oil was added first at the desired temperature, 20°C for No. 2 fuel oil and 65°C for No. 6 fuel oil. The surfactant or its components were added next. Attapulgite clay was added while stirring, which continued for an additional five minutes to obtain a gel. The coal was then added rapidly while stirring and the mixture stirred until uniform. Some of the test formulations are shown in Table 2-10. As shown in the table, the cost for stabilization ranged up to 30¢/million Btu, which was substantial compared to the present differential in coal and oil prices. Sawyer suggested that costs could be reduced by using a less expensive surfactant, less clay and a higher clay:surfactant ratio. He claims to have reduced costs to 13¢ million Btu. Some of the stabilization results are shown in Table 2-11. Only one case showed the one-month storage capability that appears to be achievable by other means. There has been some concern that the needle-shaped particles of the clay may lead to harmful asbestos-type stack emissions.

The beginning of one of the most ambitious efforts so far in raising the particulate coal content of combustible slurries has been described by Scheffee [31] of the Atlantic Research Corporation. This corporation has been aiming at a high coal concentration in gelled water and the re-

Table 2-10. Formulations of Typical Clay-Surfactant Stabilized Carbon-Oxygen Mixtures (Waring Blender Processing) [33]

Ingredients	No. 2 Fucl Oil A	B	C	No. 6 Fuel Oil D	E	F
No. 2 Fuel Oil	40%	39.7%	38.2%	—	—	—
No. 6 Fuel Oil	—	—	—	45%	44.7%	48.0%
Monazoline %	—	0.3	0.3	—	0.3	0.4
MIN-U-GEL FG[a]	—	—	1.5	—	—	1.6
200-Mesh Coal[b]	60	60.0	60.0	55	55.0	50.0
	100%	100.0%	100.0%	100%	100.0%	100.0%
Clay/Surfactant Treatment Cost ($/ton)	—	—	7.20 ($0.24/ 10^6 Btu)	—	—	9.28 ($0.29/ 10^6 Btu)

[a]Product of Floridin Company, Pittsburgh, Pa.

[b]2.8% +100 mesh; 18.2% +200 mesh; 0.8% free moisture.

sults to date indicate some feasibility for stable, fluid slurries comprised of 65–70% coal. At 65% coal, xanthan gum is used successfully as a gellant. At 70% coal, hydroxyethyl cellulose is used. Two levels of water content have been tested, either 30% or 10%, with the noncoal remainder made up by No. 6 oil. In all cases the coal has been pulverized and sieved to produce a bimodal particle size distribution so as to optimize packing density. The Bantam Mikropulverizer hammer mill has been used on the coal. Ideally, to obtain maximum filling of interstices among the large coal particles by fines, the weight ratio of coarse to fines would be 70/30, with mass mean diameters (MMD) for each class standing in the ratio 10:1. However, dissimilarity of size distributions in any lot and lack of sphericity of the particles would lead to a departure from these conditions to achieve the maximum packing density. In addition, due to limitations of pulverized coal burners, the maximum particle size must be less than 300 μ (approximately 50 mesh) so that the mill product must be classified to take out larger size coal particles, which are then recycled.

Gum stock solutions were prepared from a number of water-soluble resins, listed in Table 2-12, and slurries loaded with 60% coal of MMD about 34 μ were prepared with each of them. The naturally occurring resins (starches, guar gums etc.) lead to slurry viscosities that increased with age until the composition became essentially solid. The polyvinyl carboxylic acids, among the most cost-effective gellants, were too ion-sensitive for use and the coal settled too quickly. Best stability was shown by slurries made from carboxymethyl cellulose (CMC), hydroxy-

Table 2-11. Evaluation of Stored COM (stored at 160°F) [33]

	No. 2 Fuel Oil			No. 6 Fuel Oil		
Evaluations	A No Clay No Surfactant	B No Clay Mona.	C Clay Mona.	D No Clay No Surfactant	E No Clay Mona.	F Clay Mona.
Brookfield Visc.						
Initial—10 rpm/100 rpm	1,500/800[a]	1,000/750	7,000/880	1,200/1,000[a]	1,100/900	4,400/3,480
1 Month—10 rpm/100 rpm	—	—	11,000/2,500	—	—	8,000/4,100
Comments	Settled in 24 hours	Settled in 24 hours gummy sediment	No sediment, no SN[b] liquid in one month	Settled in 24 hours	Settled in 24 hours, gummy sediment	No sediment in one month[c]

[a]Gave very erratic readings.

[b]SN = supernatant liquor.

[c]This formulation was tested with a General Motors stability test and gave the following results after 2 weeks and 1 month at 150° F:

2 weeks:	top	44.2% coal
	bottom	49.8% coal
1 month:	top	41% coal
	bottom	59% coal

methyl cellulose (HEC) and xanthan gum. This last is the most stable for storage at elevated temperatures. Settling data for these last slurries are shown in Table 2-13, giving the effects of particle size distribution, carrier composition and coal content on stability. Settling is measured by the pendulum test of center of gravity shift. Note that, in keeping with the stability differences between 30% [32] and 40% [4,5,8], for HEC compositions settling is less at 70% than at 65%. A diagram of the pendulum settling test is shown in Figure 2-9.

Table 2-12. Summary of Water Soluble Resins Evaluated in Carriers [31]

Chemical Description	Manufacturer's Data: Trade Name	Manufacturer's Data: Manufacturer	Effective Concentration, (wt %)	Unit Cost ($1/lb)
Cornstarch, gp, industrial grade	CP3005	CPC	2–5	0.07
Waxy Maize	CP6400	CPC	2–5	0.19
Dry-Milled Corn Flour	550	Ill. Cereal	2–5	0.0571
Gelatinized Corn Flour	Ceres 820	Ill. Cereal	2–5	0.0756
Gelatinized Corn Flour	Ceredex 955	Ill. Cereal	2–5	0.0756
Modified Cornstarch	Staramic 330	Staley	2–5	0.2275
Modified Cornstarch	Staramic 747	Staley	2–5	0.2275
Modified Cornstarch	Hamaco 297	Staley	2–5	0.2650
Modified Cornstarch	Hamaco 267	Staley	2–5	0.2550
Cornstarch, gp	Pearl	Staley	2–5	0.10
Cornstarch, USP	Melojel	National	2–5	0.102
Cornstarch, USP	1551	National	2–5	0.30
Modified Waxy Maize	Capsul	National	2–5	0.50
Guar Gum	Jaguar A-40-F	Stein-Hall	0.5	0.84
Guar Gum	Jaguar A2S	Stein-Hall	1	0.84
Modified Guar	Jaguar HP-11	Stein-Hall	0.5	0.94
Modified Guar	Jaguar CMHP	Stein-Hall	0.5	1.04
Guar Gum	SC 200	Perny	0.5	0.48
Polyvinyl Carboxylic Acid Salts	Carbopol 941	B.F. Goodrich	0.2	2.65
Xanthan Gum	Kelzan	Kelco	0.2	2.63
Xanthan Gum	Blozan	Hercules	0.2	(2.60)
Hydroxyethyl Cellulose (HEC)	Natrosol 250 HHR	Hercules	0.5	1.50
Carboxymethyl Cellulose (CMC)	Cellulose Gum 9H4	Hercules	0.5	1.21
Polyvinyl Alcohol	Elvanol 71-30	Du Pont	2–5	0.675
Polyacrylamide	Cyanamev P-250	American Cyanamid	0.5–1	1.55

Table 2-13. Initial Settling Data for Selected Slurries [31]

Composition Variable				Tube ID									
				1	2	3	4	5	6	7	8	9	10
1. Coal Content (wt %)				65	65	65	65	65	65	65	65	70	70
2. Coal Composition (wt %)													
Mill Data													
rpm	*Screen*	*Sieve*	*MMD* (μ)										
3,600	27HB	—	140										55
3,600	27HB	−50	113	70	65	60	55	70	65	60	55	55	
14,000	10HB	—	26	30	35	40	45	30	35	40	45	45	45
3. Carrier Composition (wt %)													
HEC (Natrosol 250HHR)				0.5	0.5	0.5	0.5	0.25				0.5	0.5
Xanthan Gum (Kelzan)[a]									0.25	0.25	0.25		
4. CG Shift @ 2 weeks (in.)				0.096	0.098	0.077	0.051	0.066	0.054	0.039	0.034	0.045	0.063
5. Percent of Original				1.91	1.85	1.46	0.952	1.24	1.01	0.722	0.632	0.839	1.182

[a]Kelzan solution contains 0.001% Triton X-100.

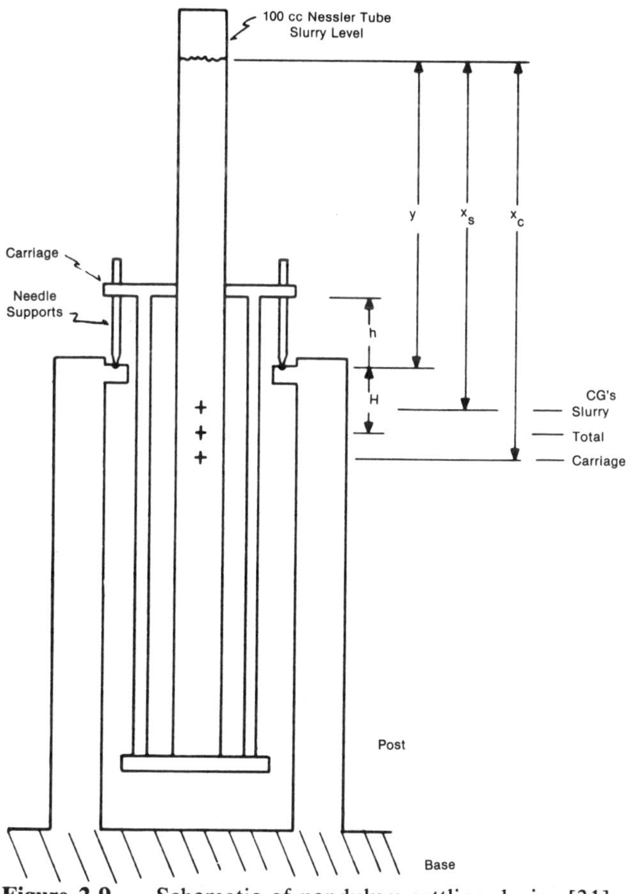

Figure 2-9. Schematic of pendulum settling device [31].

VISCOSITY AND IN-PLANT HANDLING

One of the largest questions facing COM of any coal concentration that confers stability and some chance of economic value for its use is the rapid increase in its viscosity as coal concentration rises above 30%. Not only are the flow properties of the fluid severely degraded in the linear sense of viscosity, but there is also considerable departure from linearity with respect to both rate and duration of applied shear stress. For reference, a description of viscous flow characteristics, in general, is provided in Appendix A.

The coal-oil mixtures in general exhibit to a slight degree pseudo-plastic behavior after a low critical shear threshold, and markedly increased fluidity after onset of applied shear when measured over a wide range [34]. Hence, they can be termed both Bingham and thixotropic. Manning and Taylor [4,5] recognized that to achieve stability, a gel-like structure cannot be present in the fluid medium itself. They found that pseudoplastic oil of relatively high viscosity was required. Figure 2-10 shows the contrast in viscosity behavior between their unsuitable "Persian oil" and the same oil made suitable with 0.5% sodium stearate. Viscosity for the stabilizing medium is seen to fall off slowly in the neighborhood of 1.5 poises as shear strain rate increases from 50 to 200 cm^2/sec.

Neglecting the slight departure from Newtonian characteristics, Jonnard [9] investigated the dependence of COM viscosity on coal concentration, temperature and coal particle size. As with most fluids, COM viscosity tends to fall with rising temperature. At all temperatures between 70°F and 210°F there is a rapid escalation in viscosity with coal concentration above 30%. In residual fuel oil No. 2 at 122°F, viscosity rises from 400 cp to 1400 cp as concentration increases from 30% to 40%. At 210°F the viscosity of 40% COM is only 500 cp. Since at the viscosity level of 10 poises we can expect difficulties in pumping oils through normal-sized pipes and atomizing them for combustion at the orifices of nozzles, it is clear that some heating of COM for handling

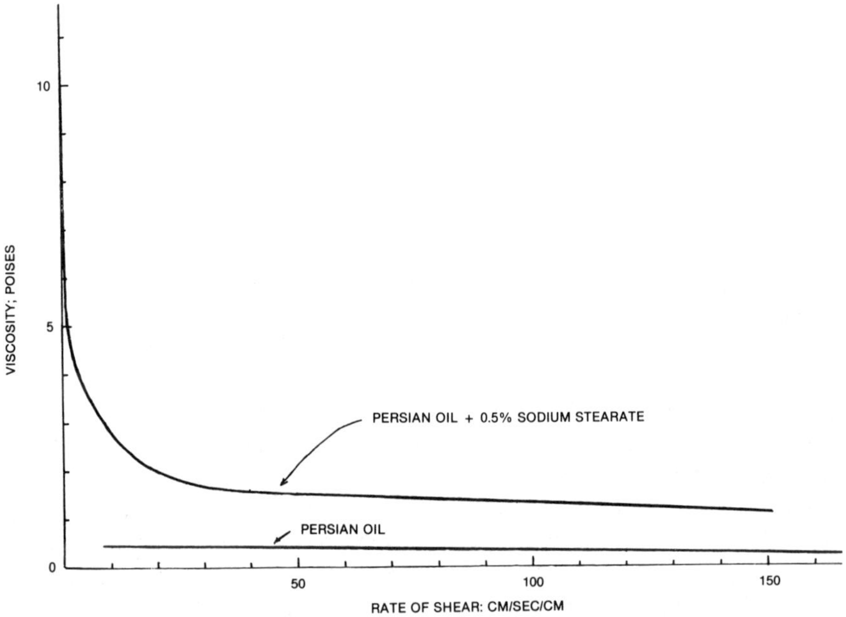

Figure 2-10. Influence of sodium stearate on viscosity in oil [4,5].

just prior to burning may be advantageous. Jonnard [9] also found that decreasing coal particle size increased the viscosity slightly. He charted the drop in fluidity (reciprocal of kinematic viscosity) with concentration as a function of temperature and found all curves converging to zero fluidity at 38% coal for both 200-mesh and 325-mesh particles.

The increase of viscosity with decreasing particle size was thoroughly explored by Rudzki et al. [12], who were interested in handling COM with particle sizes down to 1 μ. For COM of various types of coal, their data on kinematic viscosity versus concentration are shown in Figure 2-11. These data show the shift toward lower solids concentration of the vertical asymptote as the particle size drops. In a specific example of 45% coal solids at 150°F, their data in Figure 2-12 show clearly a barrier in

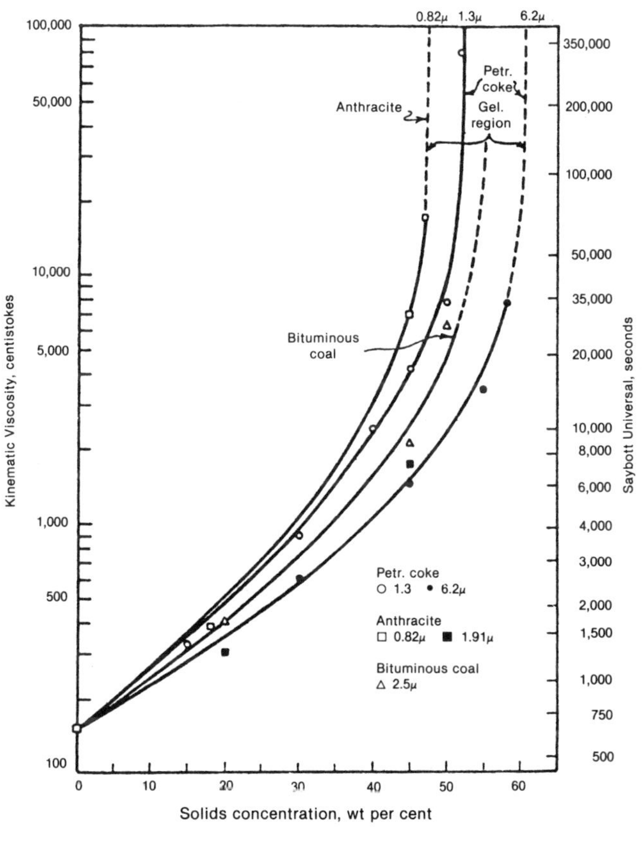

Figure 2-11. Dependence of kinematic viscosity on COM coal concentration at 150°F (65.5°C) [12].

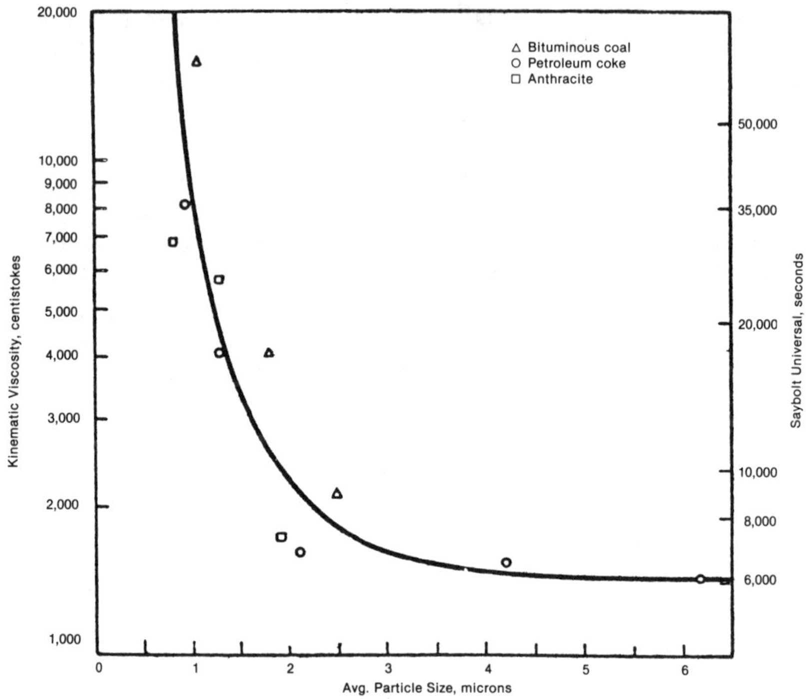

Figure 2-12. Dependence of kinematic viscosity on COM particle size at 150°F (65.5°C) and 45% coal [12].

fluidity as kinematic viscosity rises to unmeasurable values when particle size is as low as 1 μ. For particle size above 4 μ there is a very slow drop from 1600 cs with increasing particle size.

In the current effort of the DOE-sponsored projects, this early portrayal of COM viscosity behavior essentially has been confirmed with some extension work into the non-Newtonian character, especially of the more concentrated mixtures. Leffert [35], of Wayne State University, working with mixtures prepared on the GM-led consortium project, examined the dependence of viscosity on temperature. The mixture was targeted to have, by weight, 50% coal, 43.3% oil, 6.5% water and 0.2% additive. However, there was considerable spread in actual values, as shown in the data on coal concentration of Figure 2-13. These variations caused some operational difficulties, as recounted below. Viscosity temperature characteristics, examined for two specific COM samples and its constituents, are shown in Figure 2-14. They appear to be in fair agreement with Jonnard's results. Some indication of heterogeneity in the GM-prepared sample leading to added viscosity is shown by the drop in

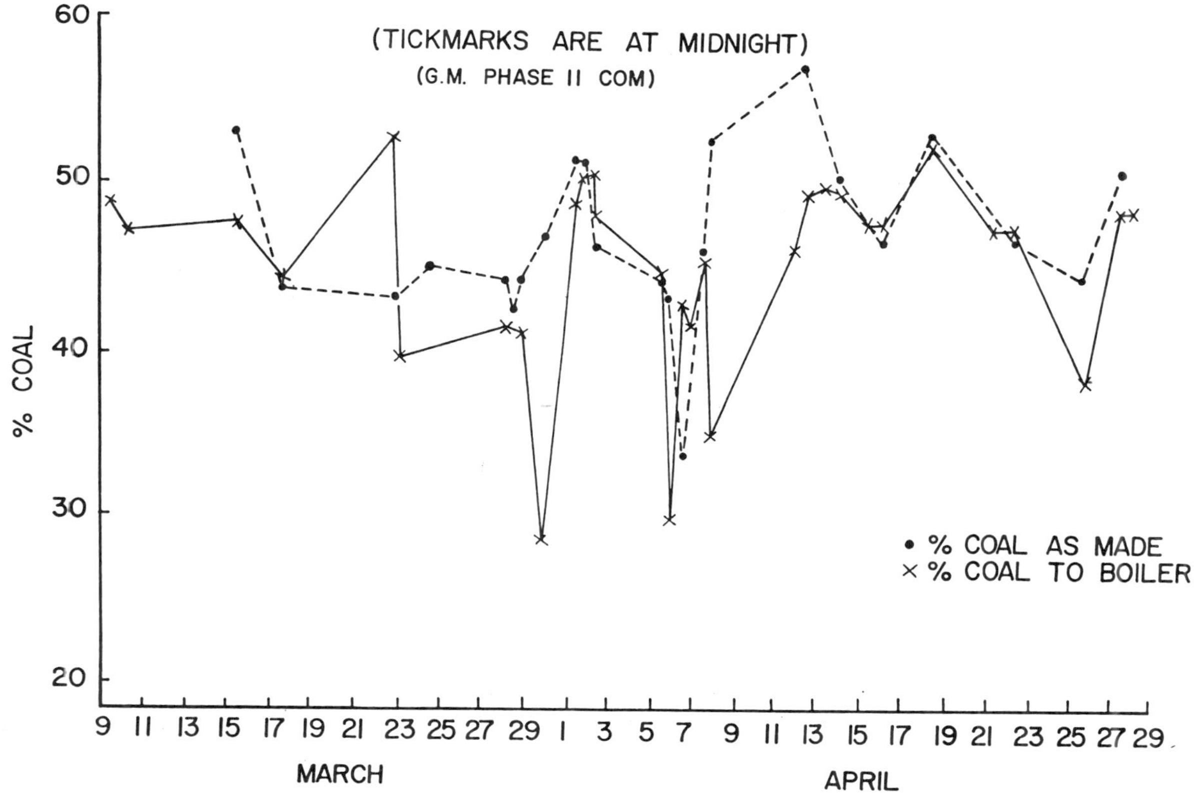

Figure 2-13. Temporal variation of coal concentration in GM COM tests [35].

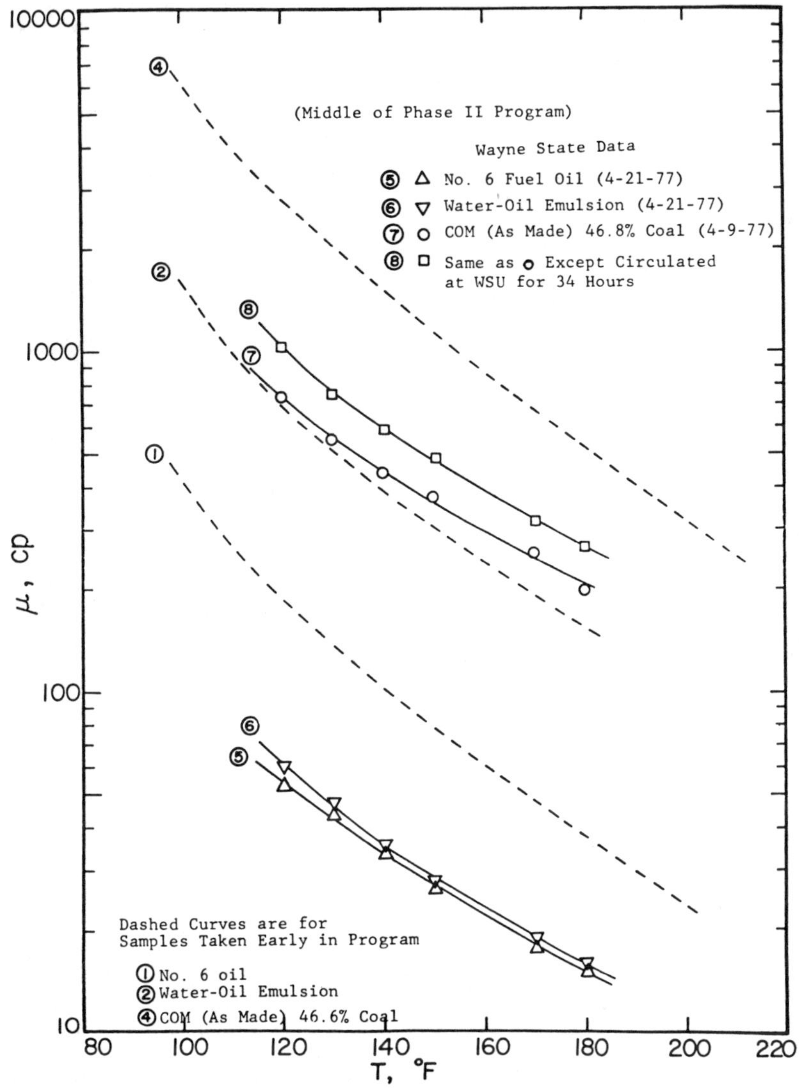

Figure 2-14. Viscosity (μ) vs temperature (T) GM COM II and its components [35].

viscosity when the mixture was recirculated in the Wayne State experiments. The fuel inconsistencies in the GM-prepared COM adversely affected combustion and other boiler operational characteristics through changes in heating value, atomization efficiency and pumping irregularities, all properties directly dependent on coal concentration and viscos-

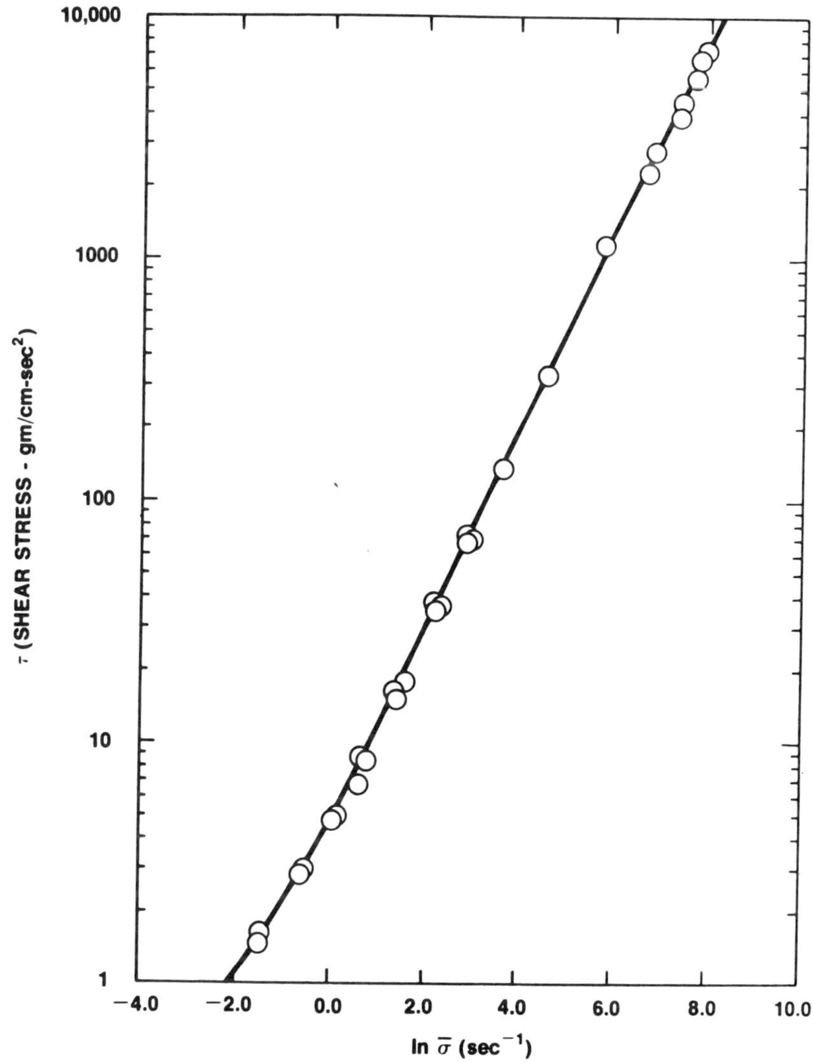

Figure 2-15. Shear stress-shear rate relationship for 40% COM (PERC No. 6 oil) (obtained with the extrusion rheometer) [32].

ity. These problems are discussed further in Chapter 3 on boiler operation using COM.

A fairly comprehensive experimental study of viscosity and flow properties was carried out by Ekmann and Bienstock [32] of the former DOE Pittsburgh Energy Research Center (PERC), recently renamed Pittsburgh Energy Technology Center (PETC). They pointed out that the

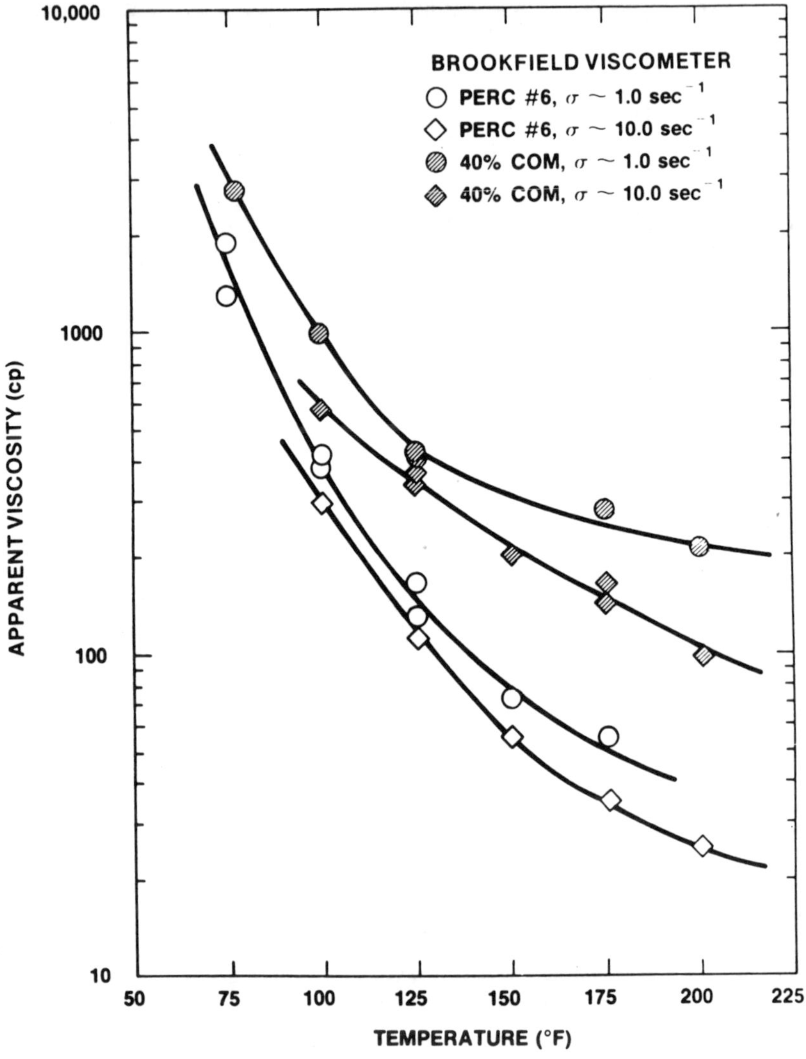

Figure 2-16. Effect of shear rate and temperature on oil and 40% COM [32].

rheological properties of COM depend both on the coal concentration and the parent oil viscosity. The addition of coal for an increase of concentration from 40% to 50% substantially increased the non-Newtonian characteristic of the mixture. Their stress-strain data for a 40% mixture are shown in Figure 2-15. The flow properties in this mixture, two oils and another 40% COM, were measured in a 1-inch schedule 40 pipe with flowrates at 210 gal/hr and temperatures at 140°F. The flowrate and inside diameter of the pipe imply an average shear rate of 35 sec^{-1}. Temperature dependence of the apparent viscosity of the flow-tested fluids is shown in Figure 2-16. Based on such viscosity behavior under assumptions of Newtonian, pseudoplastic and Bingham plastic flow properties, pressure drop per 100 feet required to maintain the volumetric flow was calculated, with the results shown in Table 2-14. Ekmann and Bienstock [32] considered the pseudoplastic properties most representative of the true rheological properties as they measured them. The PERC No. 6 oil used is a No. 6 oil available commercially in the Pittsburgh area, a supply of which was set aside for the PERC program.

Table 2-14. Pressure Gradients for Oil and 40% COM [32]

	ΔP/L (psi/100 ft)		
Fluid	Newtonian Fluid	Pseudoplastic Fluid	Bingham Plastic Fluid
Gulf Low-Sulfur Oil	17.0	—	—
40% COM—Gulf Low-Sulfur	79.4	70.4	63.3
PERC No. 6	8.0	—	—
40% COM—PERC No. 6	27.0	16.3	16.2

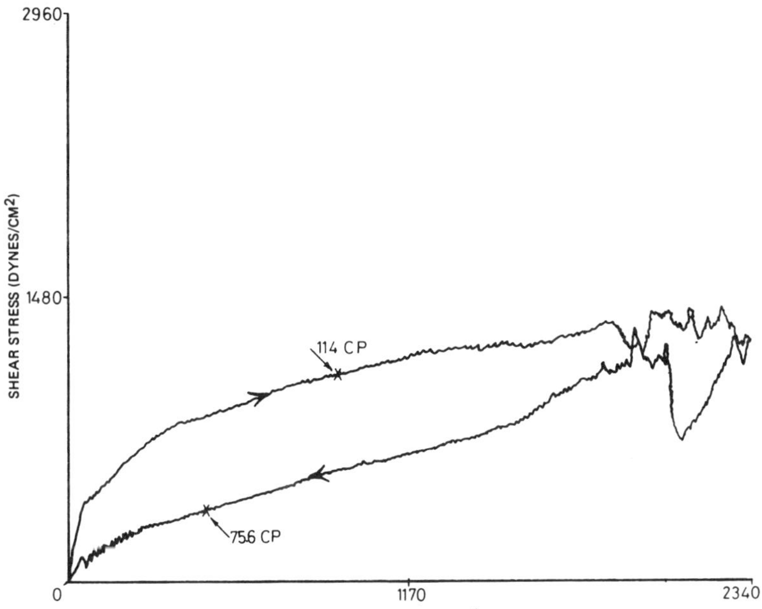

Figure 2-17. Flow curve for coal-water-oil emulsion at 40–20–40%, Penn coal, ball mill grind [25].

The non-Newtonian characteristics of coal-oil mixtures with added water were studied by Dooher et al. [25] in connection with their investigation of the combustion properties of such mixtures. They used a rotational concentric cylinder Haake Rotovisco RV3 viscometer capable of cycling through a wide range of shear strain rate, which would exhibit any non-Newtonian characteristics. COM samples were prepared from very finely ground coal, passing 90% through 325 mesh (40 μ), with a majority of particles under 10 μ. The coal, PSOC 102 from the Pennsylvania State Data Bank, is a Pennsylvania coal with a sulfur content of 1.36%. A COM of 40-20-40 coal-water-oil percentages was prepared with this coal and tested in the Rotovisco; the results are shown in Figure 2-17. There is a large separation in curves traversed with rising, then falling, applied shear at a strain rate of 1850 sec^{-1}. When the relative water content was reduced by half for coal-water-oil percentages of 44.5-11.1-44.5, this separation was reduced, as shown in Figure 2-18. In another test they found that a 40-20-40 mixture with the coarser, more frequently used coal passing 200 mesh would show even less separation and slightly lower viscosity.

In the COM R&D at Japan's Electric Power Development Company, Hatano [24] reports results on a mild concentration range of COM using a rotary-type viscometer, as shown in Figure 2-19. The same tempera-

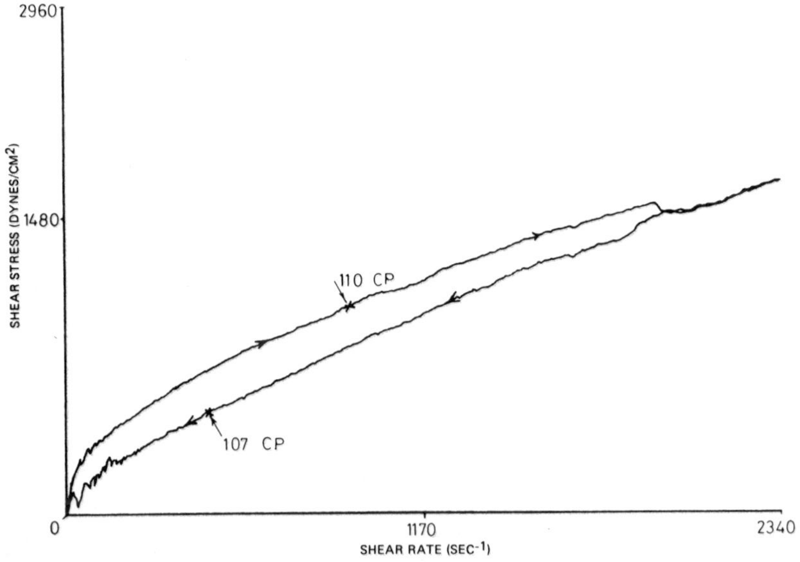

Figure 2-18. Flow curve for coal-water-oil emulsion at 44.5–11.1–44.5%, Penn coal, plate mill grind [25].

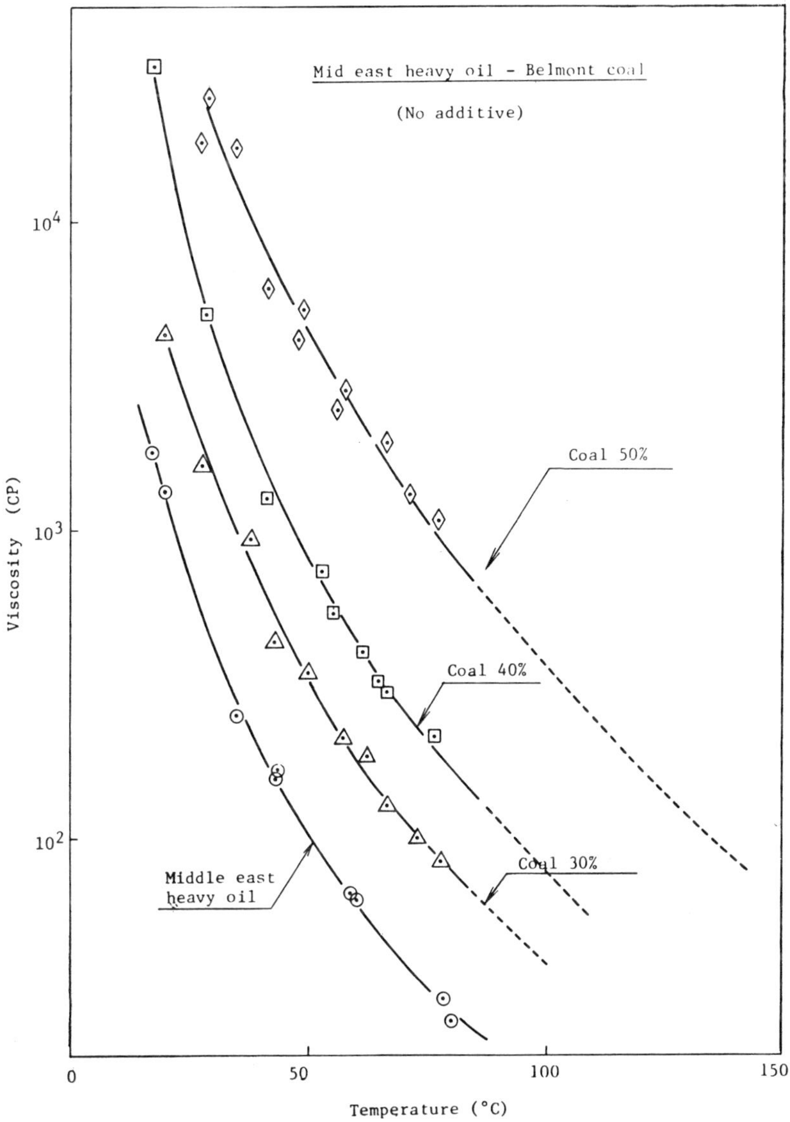

Figure 2-19. Coal:oil ratio versus viscosity (Middle East heavy oil-Belmont coal) [24].

ture dependence observed by Ekmann and Bienstock [32] is confirmed. There is also a marked viscosity increase in raising concentrations to 50%, as also observed by Ekmann and Bienstock [32] using an extension rheometer, shown in Figure 2-20.

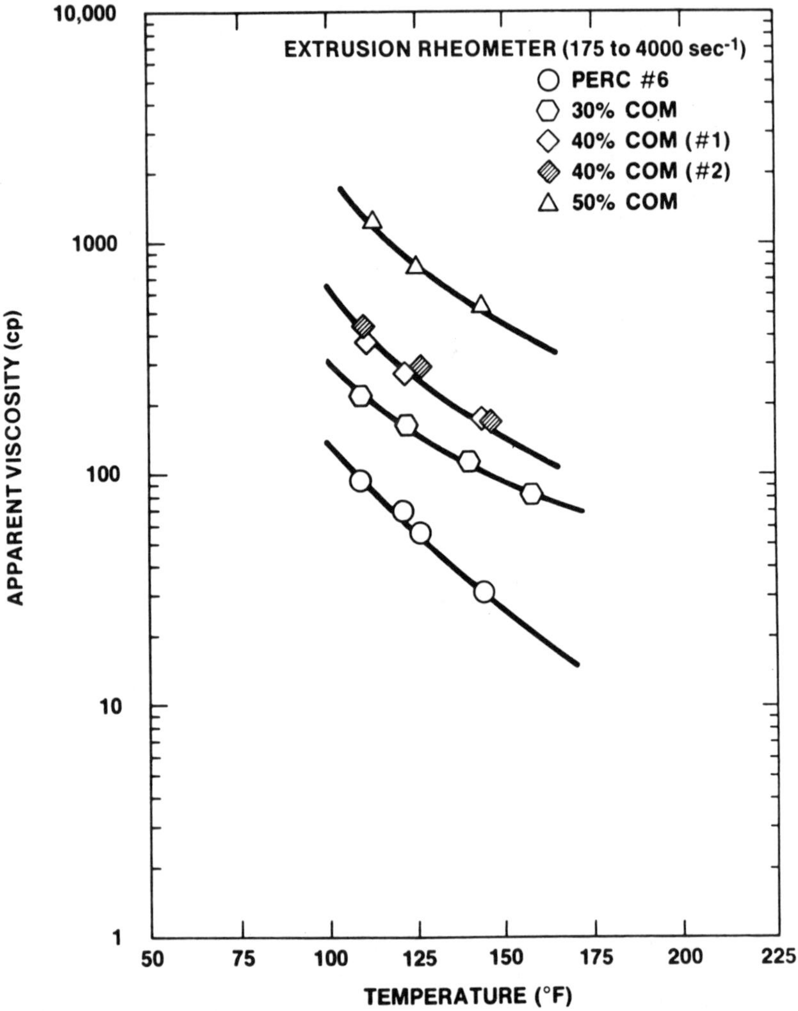

Figure 2-20. Effect of coal concentration on apparent viscosity [32].

It is quite clear that as coal concentration rises above 50%, viscosity not only rises but the non-Newtonian characteristics also become a more prominent part of the shear stress strain variation. Because the viscosity measurements do not reveal instantaneous local rheological properties, but rather some average over space and time, it is difficult to distinguish between pseudoplastic and Bingham fluid properties. A thixotropic characteristic would also contribute a pseudoplastic type of variation but it would definitively manifest thixotropicity if the strain rate is cycled with sufficient speed over a great enough range [34]. On the attapulgite clay-stabilized mixture of 50–60% coal in oil, Sawyer [3] observed a degree of apparent pseudoplasticity, which he correlated with the desirable properties for COM that led to inhibited settling at low shear and required increased fluidity at high shear conditions. Contributing factors to these rheological features are as follows:

- oil viscosity
- viscosity due to coil and clay solid particle interactions
- viscosity of clay gel

Experience indicated that clay gel is relatively unaffected by increases in temperature along with coal-coal and coal-clay interactions. The fuel oils, however, thin considerably as temperature is raised. Hence, these systems generally thin with elevated temperatures while still exhibiting pseudoplastic rheology at lower viscosity levels. The degree of pseudoplasticity can be reduced by lowering the ratio of clay to surfactant in the gel.

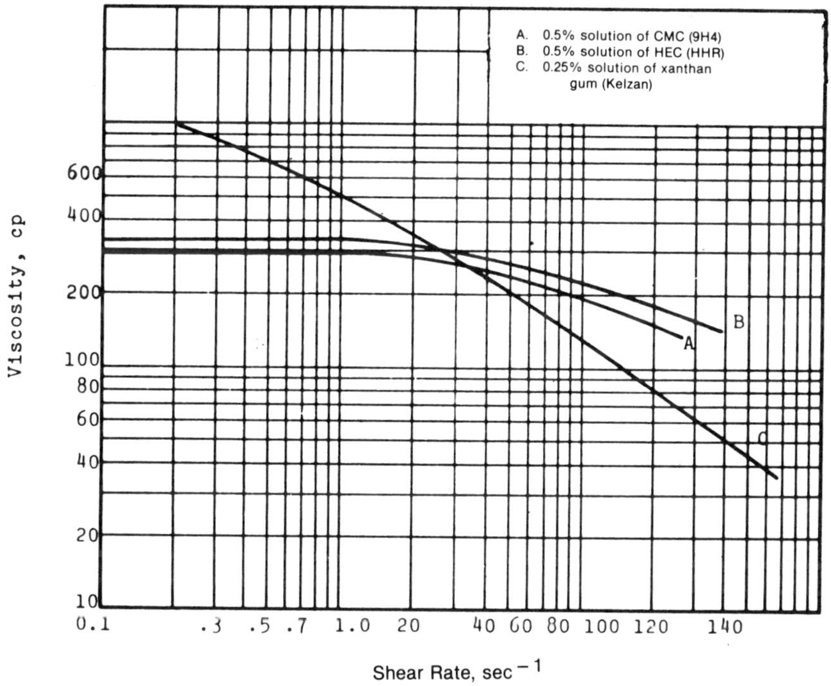

* Note: Kelzan solution contains .001% Triton X-100

Figure 2-21. Gum stock viscosities [31].

Studying the very highly loaded coal-water-oil mixtures of Atlantic Research Corp's experiments, Scheffee [31] reported very high apparent viscosities. A Brookfield RVT rotational viscometer was used for the carrier gumstock solutions, with results shown in Figure 2-21. Using the solution of lowest viscosity at high shear strain rates, 0.25% xanthan gum, the effect of particle size distribution on slurry viscosity was measured. This is shown in Figure 2-22.

From the work described, it is clearly feasible to control viscous properties of COM within a range of practical interest by methods of preparation, storage conditions, operating temperature and, where necessary, maintenance of flow conditions. The experience of the aforementioned workers has led to a present consensus in DOE that a 50–50 coal-oil concentration is a reasonable objective for current development work. Coal pulverizing to a 200- or 230-mesh level would be economical (see Chapter 4 for cost estimates) and also permit the formation of a sufficiently

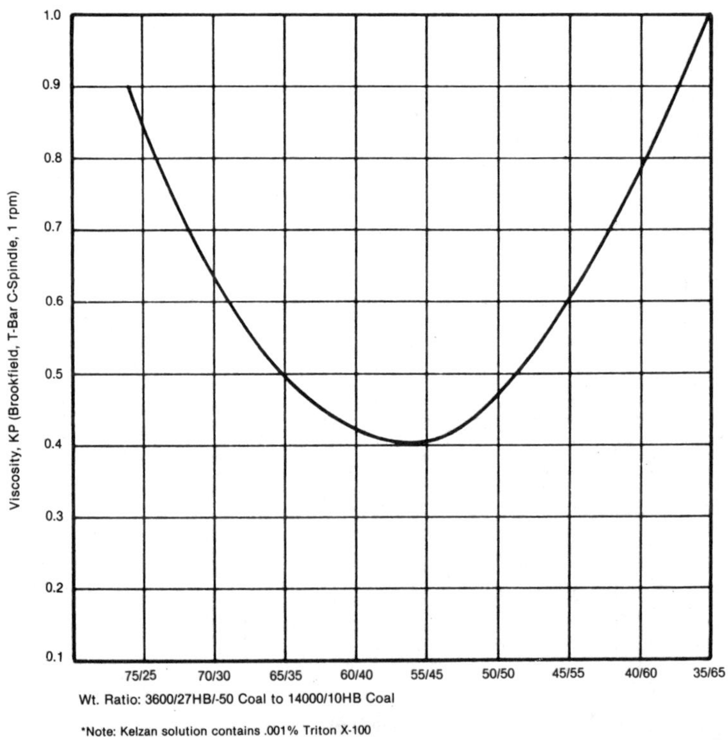

Figure 2-22. Effect of coal composition on viscosity. Coal: Bruceton bituminous; carrier: 0.25% Kelzan solution [31].

stable mixture without especially expensive additives. Better viscosity and stability properties might be attained with more finely ground coal and more expensive additives, but the extra cost may well overbalance the fuel savings at the 50% coal concentration level. On the other hand, the developments on 70% coal mixtures, especially when only 10% oil is used, promise a fuel savings level that could easily repay more costly preparation procedures.

LONG-RANGE TRANSPORT

The emphasis in the current COM work has been on an overall system in which coal would be transported as bulk-mined solids to dispersed utility or industrial users, at whose plants the coal would be ground and mixed with oil for short-term storage and subsequent use in burners. An alternative scheme of central suppliers delivering COM to users dispersed over a local service region has just begun to be studied for its economic value. Modes of transport could be intermediate-length pipelines, trucks or railway cars. The last two may appear inexpensive at first, but requirements for maintaining the coal particle suspension and mixture fluidity by continual stirring or tumbling may introduce an expensive element. The most serious work so far on long-range transport of coal slurries has been done in Canada, where the emphasis has been to deliver coal from the resource-rich western provinces to the industrial eastern provinces.

In the early 1960s, Berkowitz et al. [36], of the Alberta Research Council, carried out some pilot experiments on a 1-inch-diameter pipeline. Coal-in-oil suspensions were pumped at velocities up to 20 ft/sec (610 cm/sec) through a pipe test section almost 26 feet (794 cm) long. Velocity was measured by a venturi meter outside the test section, as shown in Figure 2-23. Such velocities indicate shear strain rates up to about 600 sec^{-1}. Their suspensions in a fairly heavy oil with viscosity of 46.7 cp at 25°C exhibited laminar flow at all velocities, and probably its flow results are most readily comparable to what one can expect for the current COM preparation. Results are shown in Figure 2-24 for coal concentration in the oil from 20% to 60%. At 60% the required pressure gradient is rather high. For a velocity of 1 fps a pressure gradient of about 1.3 feet of water (30 ft = 1 atm) per foot of line length is required. Closer studies indicated that the laminar flow was facilitated by the establishment of an axial "core" in the flow pattern, which became more pronounced with higher velocity.

Some idea of the relative economy of coal-in-oil transport can be

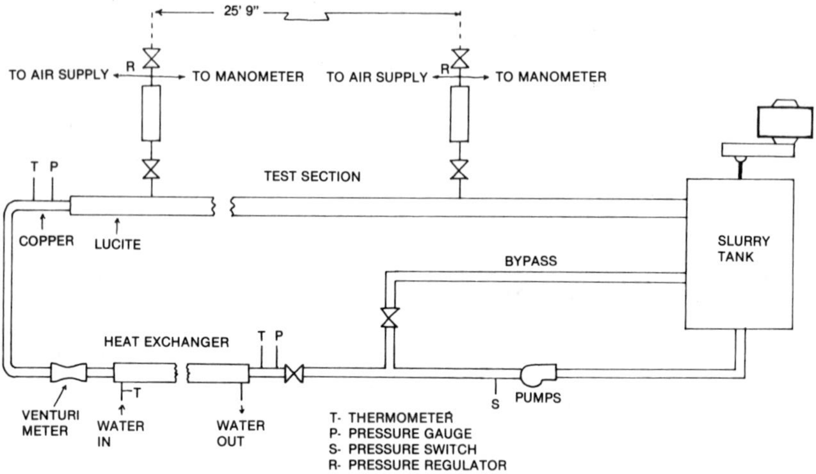

Figure 2-23. Schematic layout of experimental pipeline assembly [36].

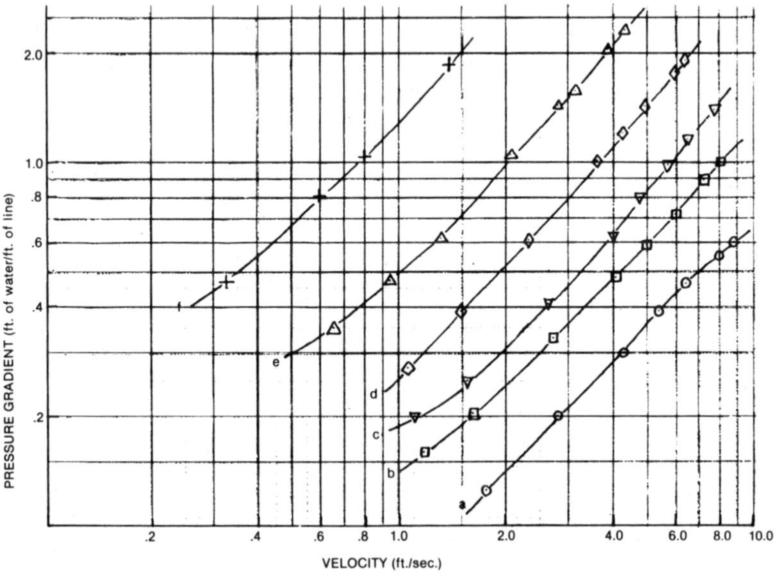

Figure 2-24. Variation of $\Delta P/\Delta L$ with flow velocity; coal A in oil No. 2: (a) clean oil, (b) 20% w/w coal, (c) 30%, (d) 40%, (e) 50%, and (f) 60% [36].

gleaned from a comparison of required pressure heat at different concentration. For example, the required pressure head for 50% coal in oil is six times that for 20% coal in oil. We therefore must balance the savings in the oil-coal price differential for 30% of the whole mixture against the expense for the extra pumping capacity. Alternatively, we can trade off extra pump capacity for a pipeline of increased diameter. Brown [37], also of the Alberta Research Council, reported his own estimate of relative economy that would be favorable because of the relatively low pipeline costs per unit of heating value. For example, for oil by pipeline, in 1972, the cost for 100 miles would have been 0.5¢/million Btu. For coal in oil up to 75% the cost would have been 2.5¢/million Btu. For the 2000-mile distance of interest to Canada at the time the economy was clearly marginal. However, for distribution in a local service area of several hundred miles in extent the economy is quite favorable, and the problems would have been mostly in preparation and end use application. In fact, Brown cited the already working operation in the U.S. for a water slurry of coarse coal particles up to 0.13 inches in diameter. The pipeline is 18 inches (46 cm) in diameter and transports 4.8 million ton/yr (4.35 million metric tons) from a coal deposit in Black Mesa, Arizona, to Southern California Edison's Mohave power plant in Nevada, about 273 miles away (506 km). The coal concentration is about 50%, with pressure gradient about 9 psi/mi (1.05 kg/cm^2/km) and design velocity of 5.8 ft/sec (1.77 meter/sec). The line uses four pumping stations passing the coal through double-acting piston pumps and has been operating since 1970. In an extensive discussion of many types of slurry pipelines, Wasp [38] cites the CONSOLIDATION line for coal in operation since 1957 and two much larger-capacity lines that were projected for operation in 1979 and 1981.

MONITORING FUEL CHARACTERISTICS

Control of combustion to supply heat at a constant rate to the working fluid of a turbogenerator is of critical importance to the operating utility, especially when we consider that for optimum efficiency the entire system is operating at the highest acceptable pressure and temperature levels. For coal-oil mixtures, combustion control is complicated by the number of physical features characterizing the mixture that influence both its flowrate through the burner nozzles and the concentrations of its combustible constituents. Concentration levels of the coal, its particle size distribution and the fluidity of the mixture through the pumps and nozzles all must be monitored and controlled to regulate combustion heat supply. This subsection discusses some of the more refined measuring techniques being tested in the current COM development work.

Particle Size in COM

One of the most important influences on the dependence of COM stability and viscosity on concentration is the particle size distribution, particularly where coal has been ground to pass a 200 mesh. Generally, the particle size distribution is inferred from the grinding and sieving procedure. Leeds and Northrup has developed instrumentation [39] to measure particle size. This feature is explicitly offered in the Leeds and Northrup MICROTRAC series. The instruments have been used briefly on the Florida Power Co. project. Using a laser source and Frauenhofer diffraction, the instrument gauges average particle size by the different patterns of coherent light scattered from the particles as expected in the arrangement of Figure 2-25, according to the relation

$$\text{Sin A} = 1.22\ \text{L}/\text{d}$$

where A is the scattering angle extent of the pattern, L is the wavelength of incident laser light, and d is the particle diameter. The instrument is automated by fitting it with rotating masks to selectively transmit to sensors the various orders of the diffraction pattern and thereby allow in situ calculation of average particle size. Some results on samples prepared in oil, water and a special nonaqueous medium are shown in Fig-

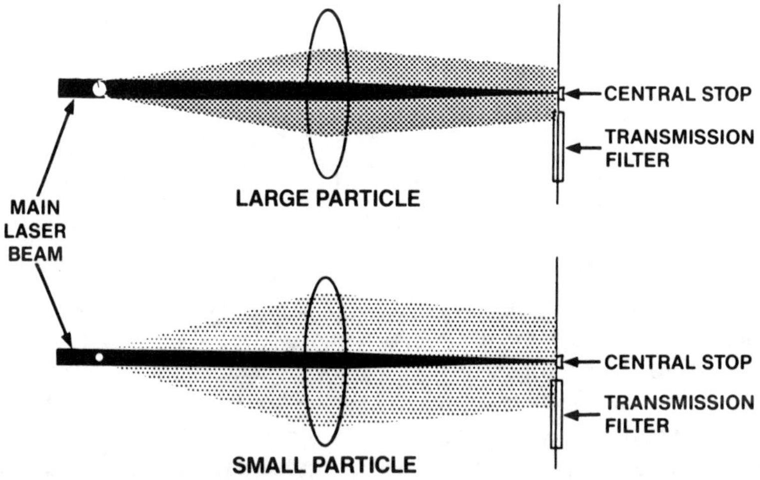

Figure 2-25. Comparison of scattered light flux angle and intensity for large and small particles [39].

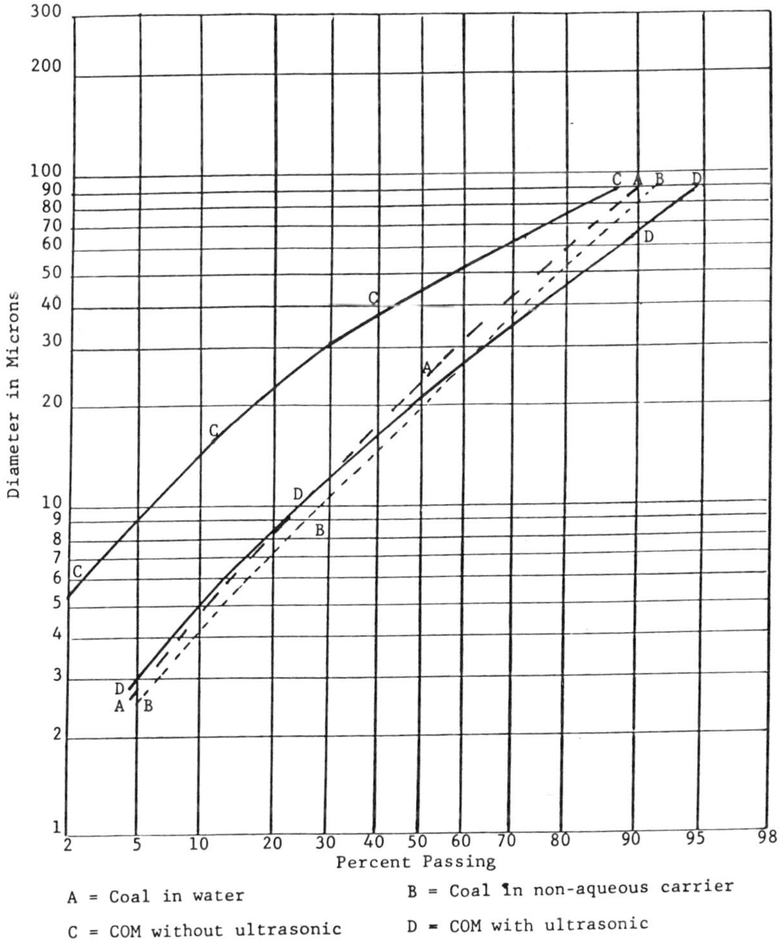

Figure 2-26. Grinding coal for COM: sieve tests [39].

ure 2-26. It was shown that the divergent sample results were due to particle clustering, which hampered passage through a 200 mesh. This was relieved by ultrasonic excitation. One serious limitation of the instrument is that the sample being observed must be diluted to 1% solids.

Sedimentation

To examine stability of COM at the Nalco Chemical Company, Kugel [40] used direct observation of samples for separate coal and oil levels. The samples were placed in glass columns and stored in an oven at 60°C

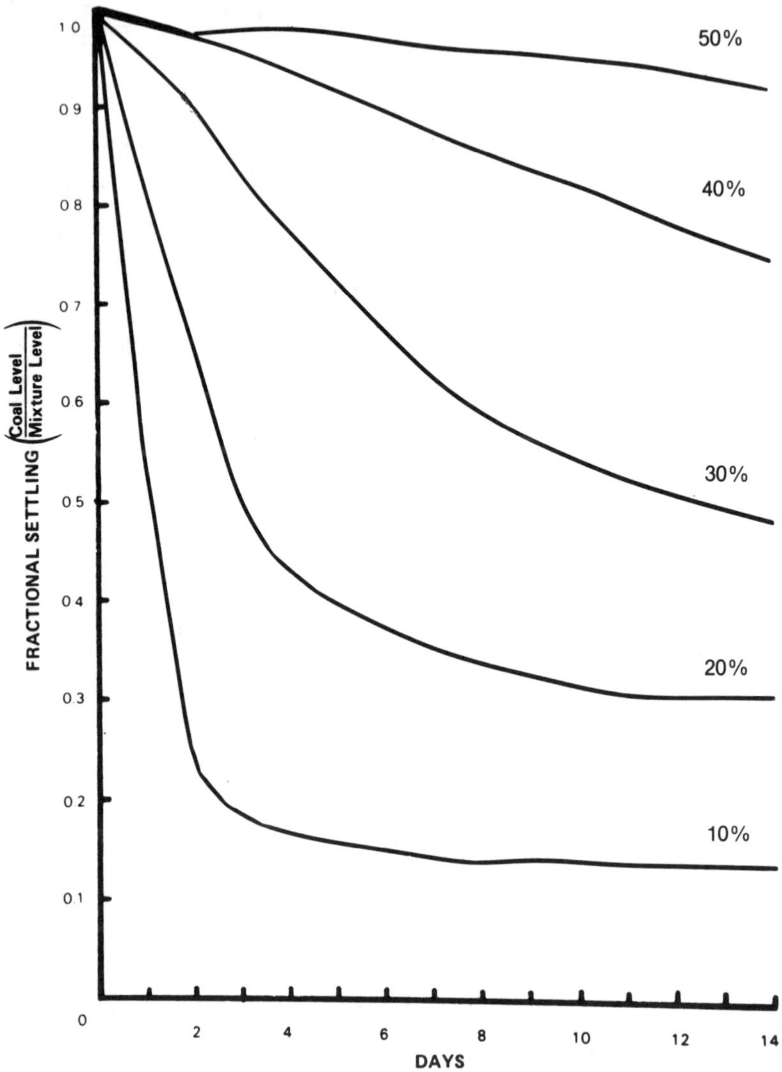

Figure 2-27. Fractional settling vs time: effect of weight percent of coal [40].

for 1-2 weeks. Sedimentation readings were taken on alternate days. Settling curves are shown in Figure 2-27. Added water generally would speed the settling at all concentrations. However, a number of patented additives were tested and showed capability of retarding this water-induced settling for the high concentrations.

Observation of settling was also carried out by Cherry and Stokes [4] of Germantown Laboratories, Inc., again in a laboratory study, using dielectric properties of heterogeneous mixtures. With further development, this kind of instrument has some promise for monitoring COM feed operations continually. It is called K-scan, and the arrangement currently tested is shown in Figure 2-28. Scanning the sample for interelectrode capacitance reveals the variation in coal concentration as a marked change in capacitance with height of electrodes. The instrument is calibrated by measuring capacitance in well-stirred mixtures of various concentrations.

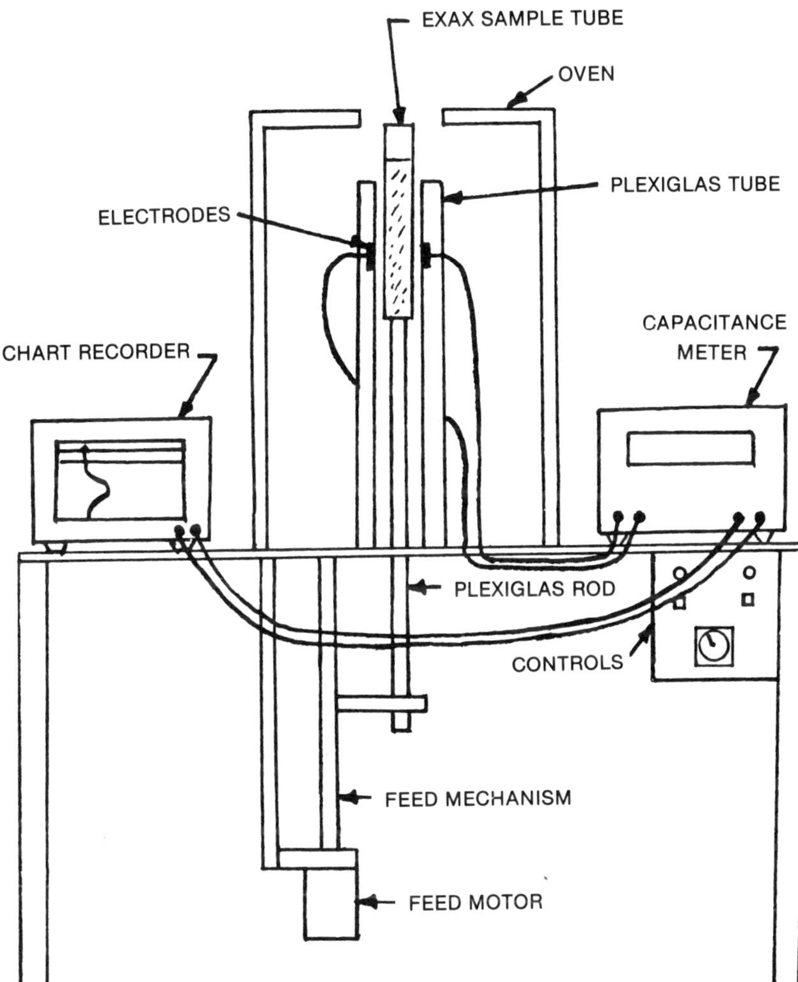

Figure 2-28. K-scan equipment schematic [41].

A fairly comprehensive study of a number of methods for examining sedimentation of COM has been carried out by Rowell et al. [42] in a joint project of the University of Massachusetts and the New England Power Service Company. They reviewed the following methods:

- Andressen pipet (side-arm sample)
- Diver method (sinking sphere)
- Hydrometer (surface density)
- Manometer (side-arm head)
- Activation analysis (neutrons)
- Photometric (turbidity)
- Sedimentation column
- Sedimentation balance

The first four methods were rejected as requiring apparatus that would be too delicate for some COM mixtures with high viscosity. The next two methods, activation and photometric, were considered too indirect. Finally they chose to rely primarily on the sedimentation column and balance.

In the first case they draw a sample from the column at given height for direct photomicrography. In this way they examine the detailed influence of additives on the coal particles. The columns were 50 cm and 1 meter in height to make the height region occupied by the sample negligible in comparison with the total column height. They found, for example, that sedimentation could be indicated by the specific gravity or volume specific weight of the sample. Figure 2-29 shows, for example, the specific volume of COM as a function of weight percent of coal, with the expectation of linear dependence clearly realized. A sedimentation balance was then set up to automate their procedure with rapid weighing of samples or testing of their density on a modified analytical balance.

More completely developed instrumentation to measure a concentration-related density in situ was reviewed by Sabadell [34] of the MITRE Corp. These would include relatively indirect measurements, such as electrical conductivity, optical refractive index, density and beam attenuation. Of these, only the beam attenuation principle seems particularly appropriate for COM. Nuclear or sonic beams, with proper calibration, seem to offer a good possibility. Sabadell [34] compared the two principles and the results are shown in Table 2-15 (a). Candidate instruments with costs are shown in Table 2-15 (b). Probably the main distinction between the two is in the radiation hazard of the gamma-ray instrument, which would militate against its use except under very well-protected conditions.

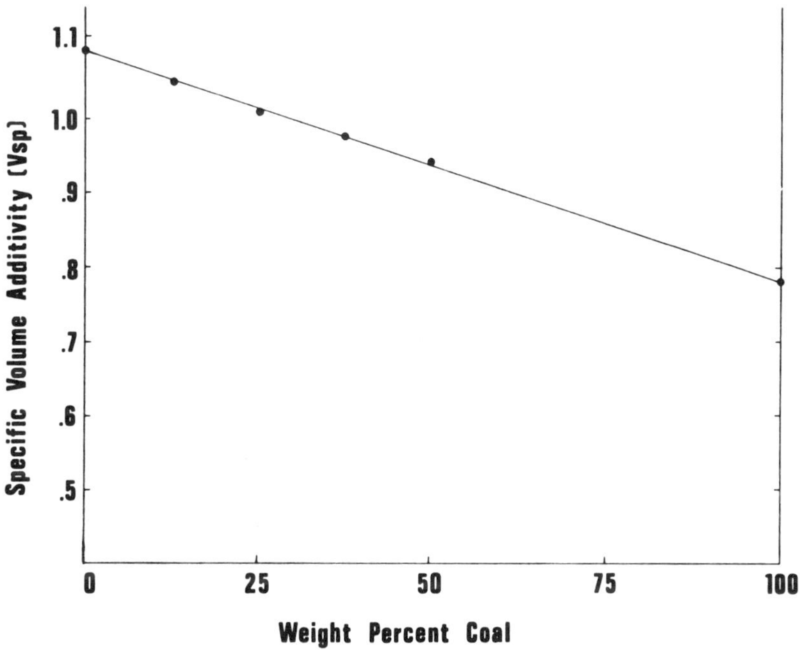

Figure 2-29. Specific volume of coal-oil mixture vs weight percent coal [42].

Fluidity

When particle size and concentration are under control we then have to manage to maintain even flow of the COM. Because of the combination of pseudoplasticity and thixotropic properties, particularly at the desirable high concentrations, this may prove the most difficult of tasks. Sabadell [34], of MITRE Corp., has surveyed instruments for continu-

Table 2-15. Measuring Coal Content in COM [34]

(a) Summary of Typical Characteristics of Potential Coal Content Instruments

Type	Data From Instrument	Range (SGU)	Accuracy (SGU)	Maximum Pressure (psig)	Maximum Temperature (°F)	Positioning	Advantages	Disadvantages
Gamma Radiation	Radiation-induced voltage	0.9–1.4	0.001	1,000	300	Insensitive	Obstructionless	Safety
Oscillational	Vibration-induced voltage	0.5–1.7	0.001	500	250	Vertical bypass installation	Simple	Dimensions and shape
Sonics	Sound velocity	0.9–1.4	0.001	3,000	300	Insensitive	Obstructionless	Process (bubbles) interferences

(b) Selected Instruments and Manufacturers for Coal Content

Principle	Price Range ($)	Manufacturers
Gamma Radiation	3000–6000	Kay-Ray, Arlington Hgts., IL Nuclear Research, Southampton, PA Ohmart, Cincinnati, OH Texas Nuclear, Austin, TX
Sonics	2500–5000	Mapco, Tulsa, OK National Sonics, Hauppage, NY

ous measurement of viscosity and found those available mainly deficient in exhibiting the thixotropic property.

There are five kinds of continuous viscometers commercially available:

- rotational
- oscillatory
- plunger
- float
- differential pressure

Of these, only the rotational type, because of its capability to isolate a sample undergoing almost uniform shear between its cylinders, can exhibit thixotropicity unambiguously and then only as manifest hysteresis due to the delayed response to applied shear stress.

The principle of this instrument operating in situ is shown in Figure 2-30. In place of the single spindle, concentric cylinders would be used for an exhaustive test of an isolated sample. The Ferranti rotational viscometer is a cone and plate type that is illustrated by the diagram and parameter table of Figure 2-31, illustrating a similar instrument by Haake [43]. Rheograms illustrating thixotropic hysteresis in shear stress-strain variation for 50% COM at 100°F (38°) and 50°F (47°C) are shown in Figure 2-32. The curves were each traced in a cycle of 4 minutes on a Ferranti viscometer. Two minutes were used to reach maximum shear stress. Even in coal-free No. 6 oil, thixotropic behavior is shown in temperatures as high as 100°F (38°C), although not apparent at 150°F (47°C). For all these cases, thixotropic hysteresis in a short cycle is much more pronounced than the 40-20-40 mixture results of Dooher et al. [25] shown in Figures 2-20 and 2-21. Sabadell's [34] results surprisingly indicate no faster relaxation at the higher temperature. Such differences with temperature and concentration, using available instrumentation, may be studied by varying the cycle time. A comparative survey of the properties of various kinds of viscometers and a list of some of the leading manufacturers is provided in Table 2-16.

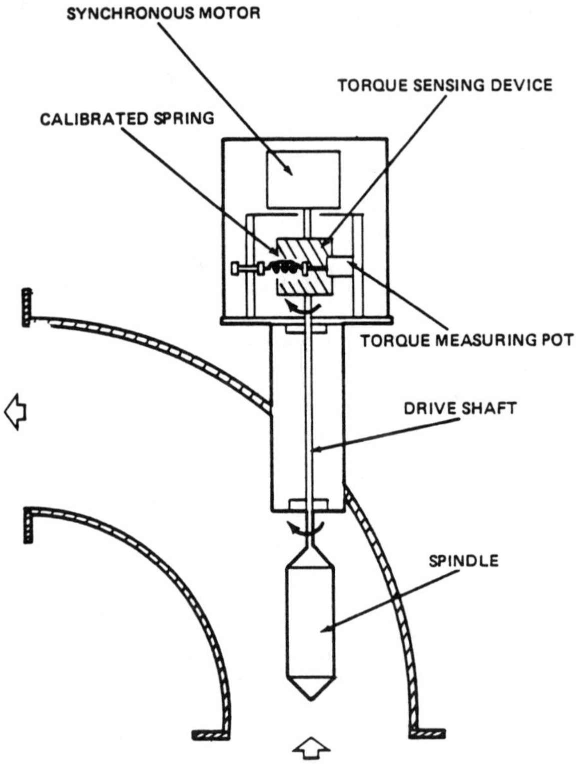

Figure 2-30. Schematic of rotational-type viscometer [34].

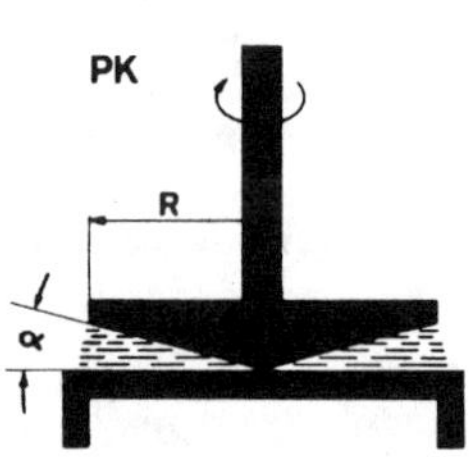

Sensor System	PK I, 1°	PK I, 0,3°	PK II, 0,3°
Recommended viscosity range (mPa · s)	100– $5 \cdot 10^4$	100– $4 \cdot 10^4$	500– 10^5
Maximum viscosity range (mPa · s)	70–10^5	50–10^5	200– 10^6
Cone angle (10^{-2} rad)	1.47	0.524	0.524
Rotor radius R_i (mm)	14	14	10
Material	steel		
Sample volume (cm^3)	0.1	0.1	0.1
Temperature range (°C)	0/100	0/100	0/100
Reproducibility (±%)	2	2	3
Comparability (±%)	4	4	6

Figure 2-31. Haake viscometer sensor system, PK [43].

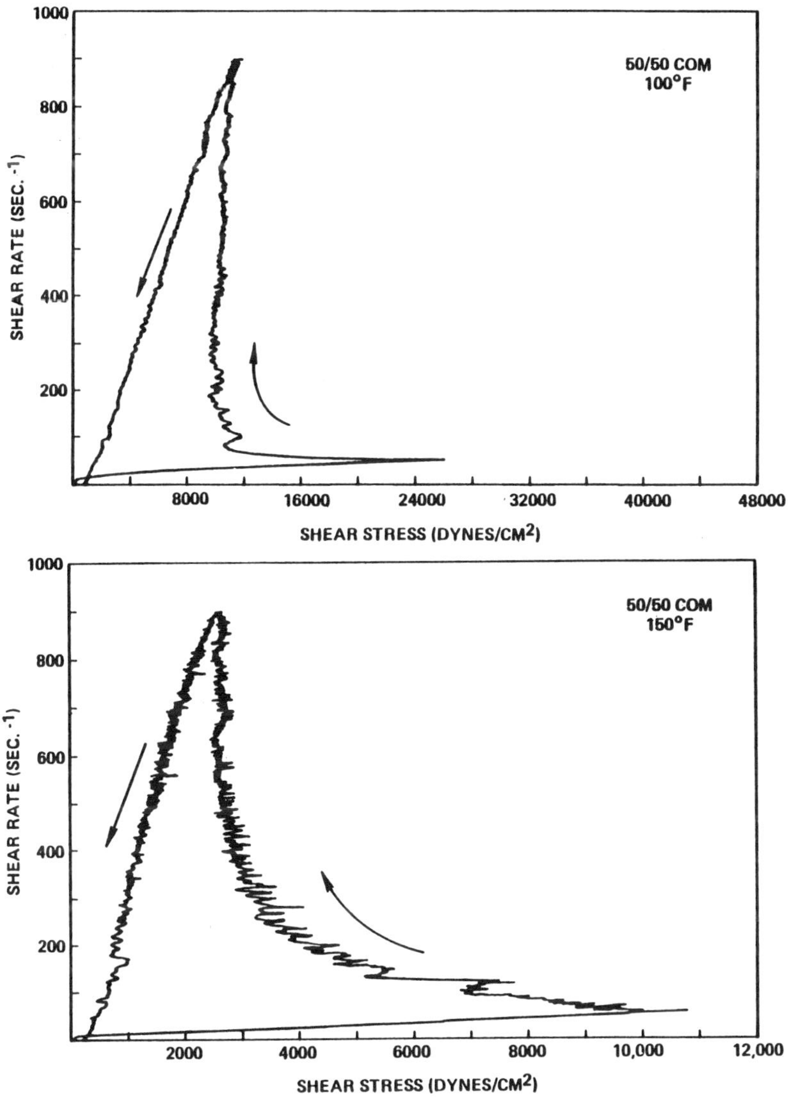

Figure 2-32. Manifestation of COM thixotropicity in Ferranti rheograms [34].

Table 2-16. Measuring Viscosity of COM

Characteristic	Weighing Factor (1-10)	Rating (1-5)					
		Rotational		Oscillatory		Plunger	
		Rating	× Weight	Rating	× Weight	Rating	× Weight
(a) Comparison of Three Methods of Measuring COM's Viscosity [34]							
Reliability/Maintenance	10	3	30	3	30	2	20
Range	9	5	45	2	18	4	36
Shear Rate Versatility	8	5	40	1	8	2	16
Accuracy/Repeatability	8	4	32	2	16	3	24
Operational Simplicity	8	4	32	3	24	3	24
Operating Conditions (temperature, pressure, aeration, etc.)	7	2	14	4	28	3	21
Response Time	6	5	30	4	24	2	12
Installation	5	2	10	5	25	2	10
Safety	4	3	12	4	16	3	12
Score			245		189		175
Normalized to 100			100		77		71

(b) Selected Instruments

Operating Principle	Price Range ($)	Manufacturers
Rotational	2500–6000	Brabender, Hackensack, NJ Brookfield, Stoughton, MA Eur-Control, Decatur, GA Ferranti, Plainview, NY Haake, Saddlebrook, NJ Hallikainen, Berkeley, CA
Oscillatory	2400–4000	Automation Products, Houston, TX Bendix, Baltimore, MD Nameter, Editor, NJ
Plunger	3000–4000	Gam-Rad, Detroit, MI Norcross, Newton, MA

CHAPTER 3

COAL-OIL MIXTURE COMBUSTION CHARACTERISTICS AND PROBLEMS

With feasibility established for preparing stable COM fuel at temperatures that assure sufficient fluidity to allow conventional pumping apparatus, the completion of a COM boiler system has no serious barriers by the standards of prior work. The reports of the 1930s and 1940s [4,5,8,9] described how little difference there was between burning COM and oil. To assure sufficient economy to establish an enduring practice, the current DOE project aims at somewhat higher coal concentrations. These will probably entail some difficulties in handling for the boiler system and in combustion products, which must be controlled to safeguard the

Table 3-1. Technical Issues in DOE COM Project for Combustion and Pollution Control [18]

System Function	Approaches/Issues
Heating (for atomization)	High-temperature COM stability
Burner	Nozzle erosion
	Burner types
Furnace	Combustion efficiency
	Flame size and shape
	Slagging
Heat Transfer Surfaces	Ash deposition and quality
	High temperature corrosion and erosion
	Soot blower performance
Overall Boiler	Derating
	Efficiency
	Reliability/maintainability
Fly Ash	Particle size distribution
	Fly ash resistivity and ESP performance
	Fly ash characteristics and baghouse performance
SO_x	Clean fuel
	Low-cost FGD
NO_x	Higher fuel nitrogen

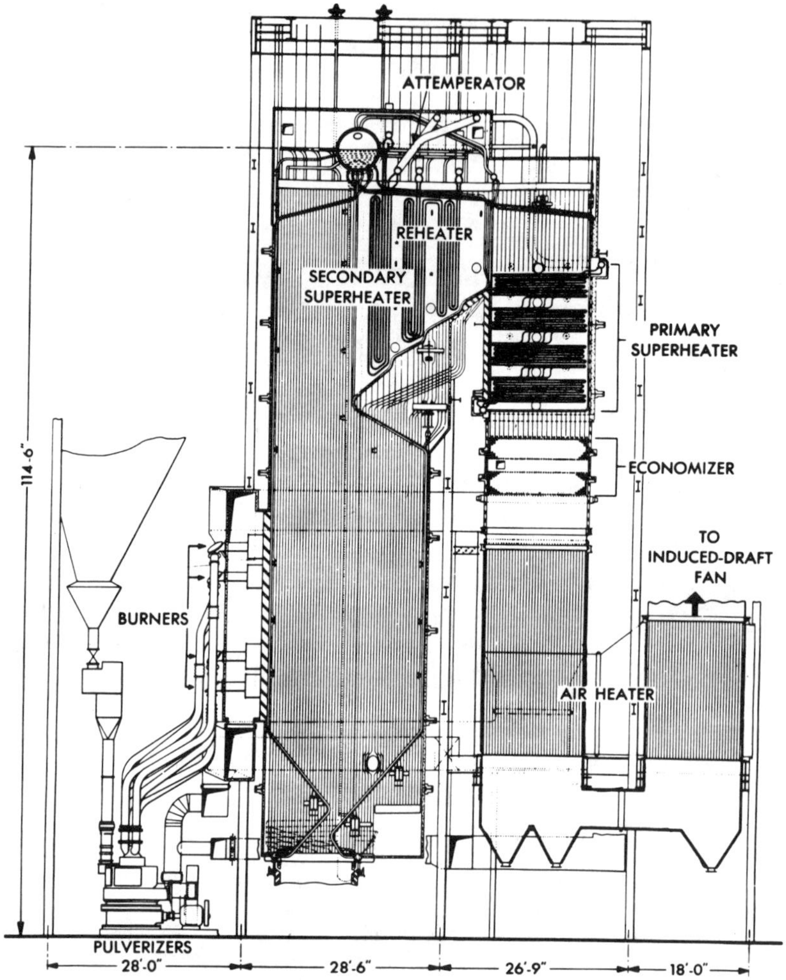

Figure 3-1. NEPSCO Salem Harbor unit No. 1 boiler [45].

combustion and heat exchange equipment and the local environment. The major technical divisions of these concerns, as outlined by DOE for their program, are shown in Table 3-1.

A typical installation for which these concerns apply is the subject test boiler of the New England Power Service Company Project (NEPSCO), described by Dunn [44] and Batra and Tsai [45], as illustrated in Figure 3-1. It is a Babcock and Wilcox front-fired radiant boiler rated at a steam flow of 625,000 lb/hr with a superheat/reheat temperature of 1000°F and a design pressure of 1675 psig. This pressure is somewhat lower than

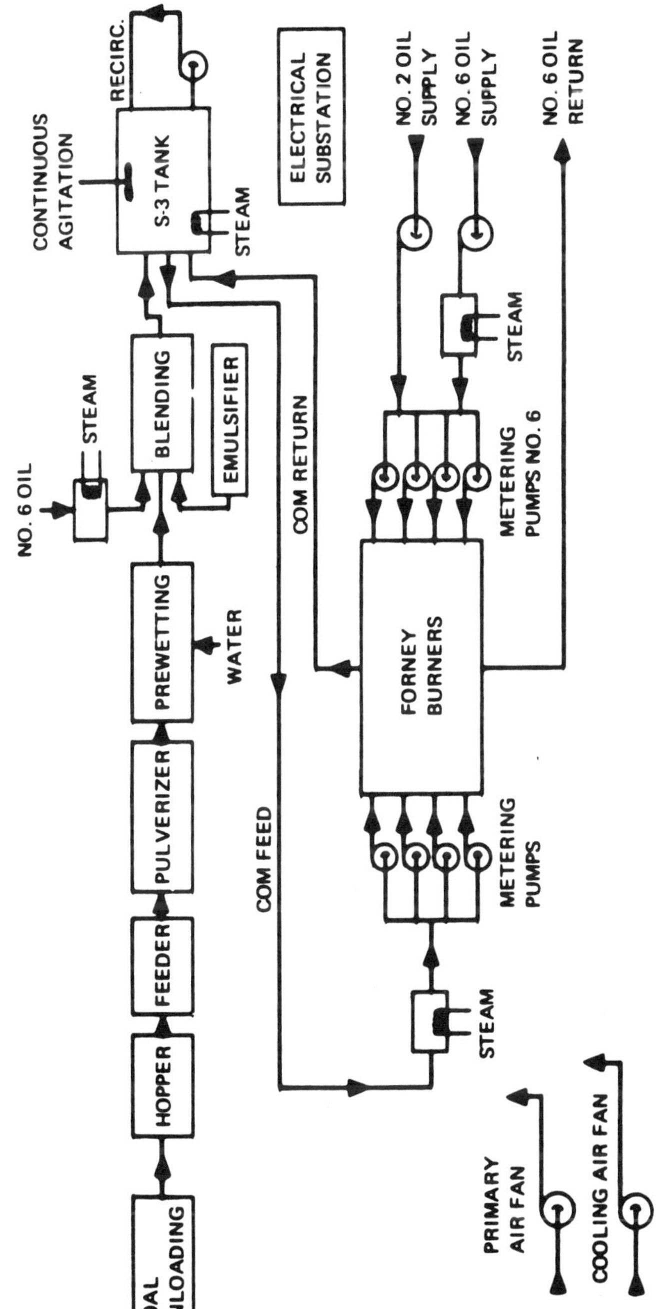

Figure 3-2. Simplified flow diagram for combustion of coal-oil mixture [44].

the upper range of U.S. practice, which has extended to over 3000 psig in steam boilers. For COM, most of the modification will take place in the leftmost segment of the diagram devoted to pulverizers and burners. These are being reworked to arrive at a system that will operate according to the flow diagram of Figure 3-2 [44]. From this diagram we see that the principal changes are the introduction of the coal-oil mixing apparatus and the substitution of Forney burners for proper atomization and combustion of the COM. This section treats the following five technical areas, embracing the DOE program and sequenced according to the progress of the COM fuel through the boiler:

- Fuel heating, flow and atomization
- Burners and combustion
- Combustion products and in-plant effects
- Heat exchangers and boiler characteristics
- Pollution control

FUEL HEATING, FLOW AND ATOMIZATION

The NEPSCO project is attempting a combined heat and flow operation for relatively low-coal-concentration COM (30:70). In Figure 3-2 the introduction of steam as a heat source for the oil and the COM is indicated. An experimental method of injection-type mixing with an "eductor," illustrated in Figure 3-3a, has been described by Batra and Tsai [45]. Pulverized coal is mixed into a pressurized stream of oil by differential pressure across the powdered coal. The pressure energy of the oil is converted to momentum of a jet of oil droplets by the converging nozzle of the eductor. This jet entrains the pulverized coal in the hopper attached to the throat of the nozzle. Mixing between coal powder and oil takes place as the COM is discharged from the eductor. A more conventional method planned for the full-scale tests is shown in Figure 3-3b. Steam is circulated through coils in the oil tank to maintain required temperature for fluidity.

Another example of the use of steam to preheat the COM for introduction to pumping and burner nozzles is described by Demeter et al. [3] for the Pittsburgh Energy Research Center project. A simplified flow diagram is shown in Figure 3-4. Steam is used for heating in fuel oil storage and both the mixing and feed tanks for the COM. Mixing is done in a round bottom tank about 1.6 meters in diameter with a three-bladed marine propeller-type mixer, gear-driven at a speed of 175 rpm. There are two propellers at 15 cm and 75 cm from the bottom of the tank along

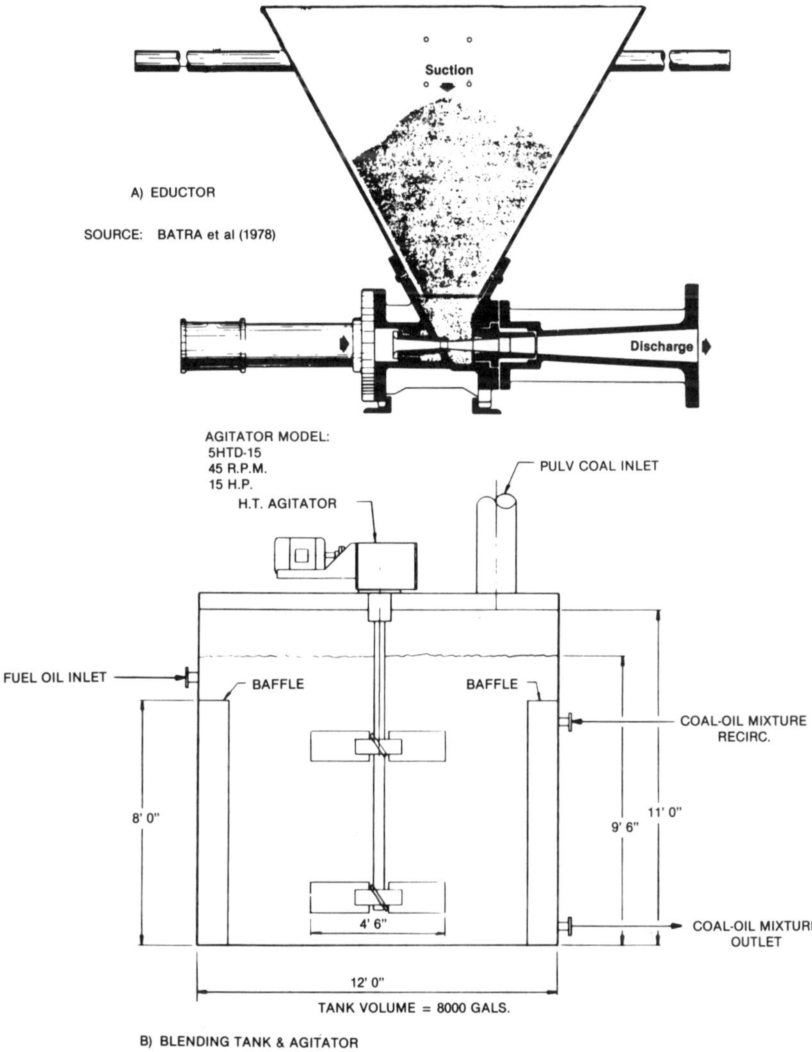

Figure 3-3. NEPSCO COM preparation.

the centerline to establish a good vortex to receive the pulverized coal. The height of the fluid above the propeller influenced the vortex and mixing, and it was therefore limited to about 45 cm above the upper propeller.

A slurry of 20% pulverized coal, nominally 90% through 200 mesh in No. 6 fuel oil, was fed to the burner via a Moyno pump and, later, a Viking gear pump. The atomizing air in the burner had to be adjusted to

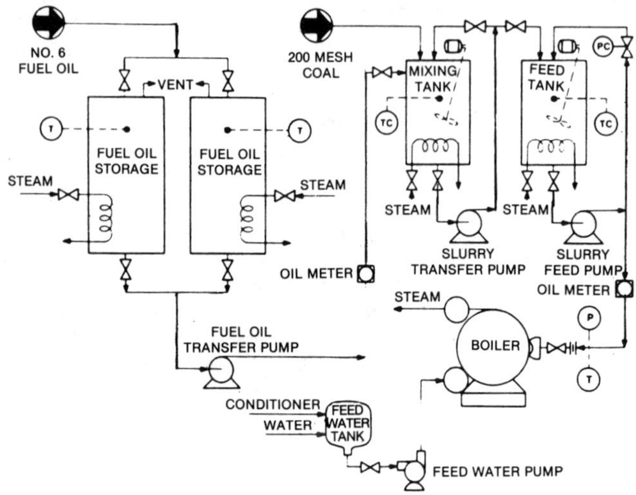

Figure 3-4. Flow diagram for the Perr firetube boiler COM test facility [3].

a 30% excess air level to promote a carbon burnout rate sufficient to curb carbon deposition on the refractory of the combustion chamber. After only 380 hours of a 1000-hour slurry combustion test, operation was stopped by coal deposits in the fuel metering pump, an auxiliary electric heater and the solenoid valve at the burner. When slurry flow through the heater was interrupted, coal settled out and deposits formed in the heater. When the solenoid opened for slurry flow to the burner nozzle, these deposits broke loose and spasmodically clogged passages in the solenoid and nozzle to eventually extinguish the flame.

The problem was alleviated with a modified slurry feed shown in detail just prior to the burner in Figure 3-5. Heaters were placed in the mainstream fuel transfer line rather than in the burner line to flush them with greater slurry volume. Strainers were installed to remove oversized coal particles, and back-pressure regulators improved pressure control. Final input was controlled by maintaining a given static pressure at the burner nozzle after the oil metering pump was removed.

The more concentrated COM at Accurex has been tested on a simulated boiler by Busch and Brown [46], who used much the same arrangement and preheated entirely by electric resistance dissipation. Coals, pulverized to 70% passage through 200 mesh, were mixed with fuel oils at a 30:70 ratio in a high-turbulence batch mixer supplied by Littleford Brothers. A proprietary suspension additive by Carbonyl Co. as a 5% aqueous solution was added to ensure a homogeneous mixture. It constituted 3.75% of the total COM weight. The COM was stored in drums at ambient temperatures of 10–15°C for 21 days before firing. About

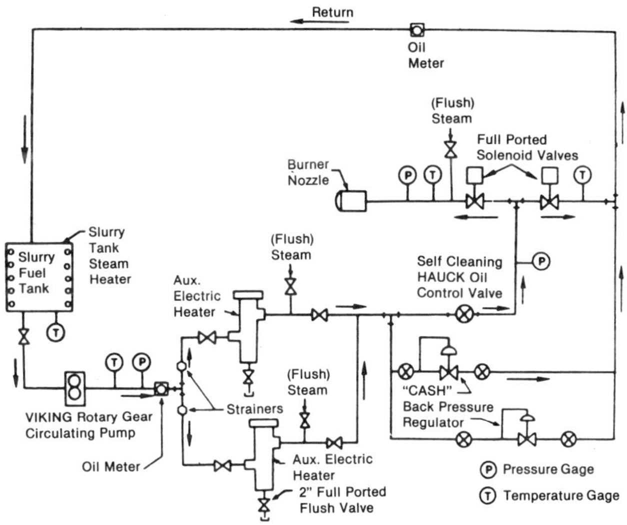

Figure 3-5. Firetube boiler slurry circulating loop at end of test [3].

4 hours prior to use, the drums were heated and mixed with a propeller and recirculation pump until a homogeneous mixture temperature of 150°–170°F (47°–58°C) was achieved. This mixture was then pumped to the delivery system shown in Figure 3-6. It includes a 120-gallon heated storage tank with a propeller recirculation mixing system in which the fuel is kept at 120°–180°F (49°–82°C). The mixture was delivered to a burner using a Viking C-32 rotary pump with varispeed control.

In laboratory tests at the Electric Power Development Company in Japan, a number of homogenizers or mixers for COM were tried [24]. Their characteristics are shown in Table 3-2. The most successful of these was the TKL Line Homomixer for its combination of relatively low required input power, low degree of separation and high-temperature homogeneity. It is illustrated in Figure 3-7; Table 3-3 indicates its test results. Mixing and pumping to burners was tried at coal concentrations of up to 55%. Hatano felt that 50% was a practical limit. When heated at 100°–150°C, the viscosity was 200–300 cp, with the higher viscosity causing some trouble in the burner atomizers. Their results on viscosity versus concentration at 70°C for a number of different oils and coals, using their mixing equipment, are shown in Figure 3-8.

The program at Accurex that has been developing equipment for the highest practicable COM coal concentrations is still in its initial stages, as reported by Shimizu and Derbidge [47]. Their COM is discharged from a ball-mill trommel designed to remove tramp oversize (i.e., tramp

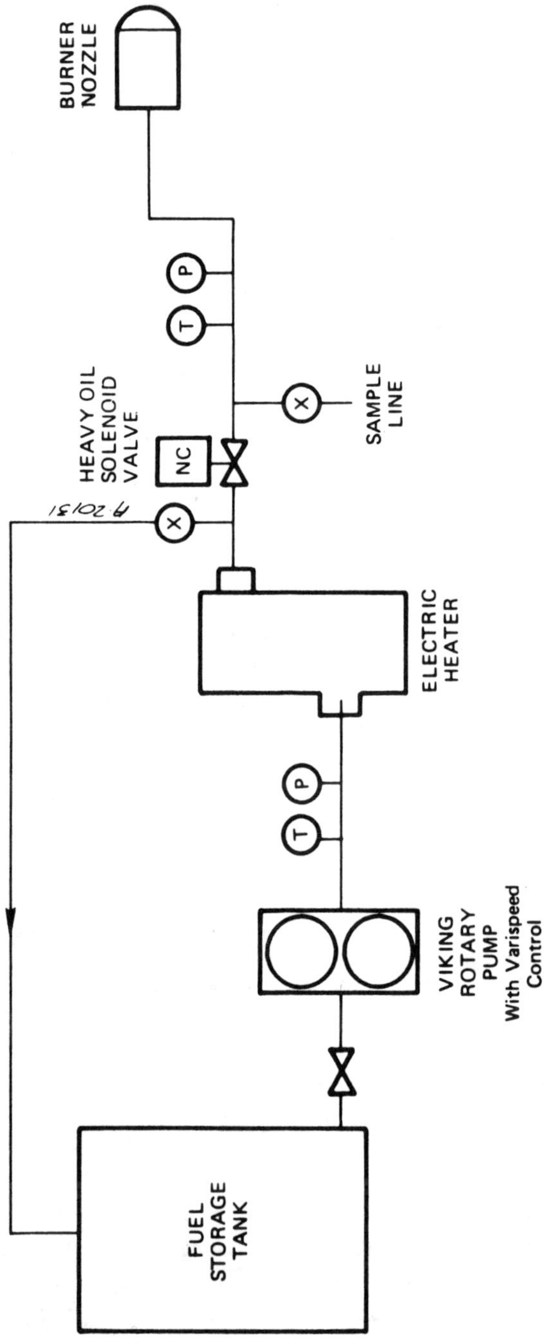

Figure 3-6. Accurex coal-oil delivery system [46].

Table 3-2. Table of Mixer Comparison [24]

Items		Disper Mill Hosokawa Iron Works	TK Line Homomixer Tokushukikakogyo	Flow Jet Mixer Funken Co. Ltd
Capacity	(kg/hr)	548	351	800
Input power	(kwh/t)	15.6	0.9	3.7
Temperature				
Inlet	(°C)	38	72	60
Outlet	(°C)	68	70	40
Viscosity				
Right after production	(cp) at 70°C	1390	1820	980
Upper part 4 days after production	(cp) at 50°C	1910	1330	170
Bottom part 4 days after production	(cp) at 70°C	1440	2230	2940
Degree of separation[a] 4 days after production	(%)	1	9	74
Rod penetration time 7 days after production	(sec)	0.5	4	7
15 days after production	(sec)	0.5	13	25

$$\text{Degree of separation} = \frac{Di - Dv}{Di - Do} \times 100$$

Di = COM specific gravity of right after production.
Dv = COM specific gravity of upper part after 4 days.
Do = Specific gravity of oil.

and iron ball chips), as shown in Figure 3-9. Heated oil is fed to the ball mill to introduce the mixture during the grinding process with the temperatures controlled through a steam-fired, finned-tube heat exchanger. The COM is collected in a pump from which it is conveyed to a control tank using lobe pumps modified for the purpose by the Tuthill Corporation. Only COM meeting both density and grind requirements is passed to the subsystem preceding the burners. The main factors driving this subsystem design are:

- erosion
- viscosity
- plugging and settling

To prevent erosion, the lobe design uses a low-speed rotation (84 rpm) and avoids metal-to-metal contact. Erosion in control valves will be min-

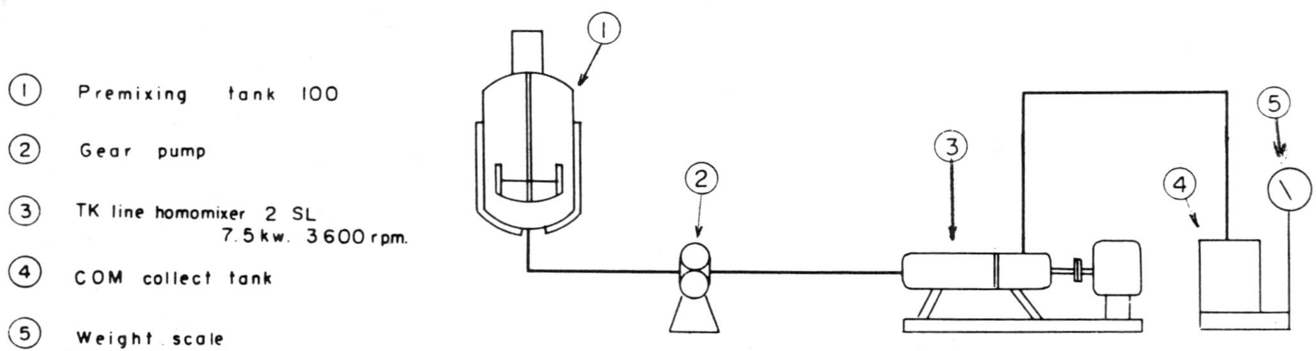

Figure 3-7. Japan Electric Power Development Co. mixing system for COM [24].

Table 3-3. Test Results of TK Line Homo Mixer [24]

No.	Passing times	Temperature		Mass flow	Cumulative Consuming Power	Viscosity			Degree of Separation	Rod Penetration Time	
						Right After Production	Upper Part 4 Days After Production	Bottom Part 4 Days After Production	4 Days After Production	7 Days After Production	15 Days After Production
	—	Inlet (°C)	Outlet (°C)	(kg/hr)	(kWh/ton)	(cp.)	(cp.)	(cp.)	(%)	(sec.)	(sec.)
1	0	70		—	—	1340	1070	2030	10	5	20
2	1	75	75	322	0.47	1780	1130	2560	6	5	13
3	2	73	71	305	0.66	1640	950	2240	15	3	11
4	3	69	70	321	0.70	1330	1050	2080	13	3	16
5	4	72	70	351	0.90	1820	1300	2230	9	4	13

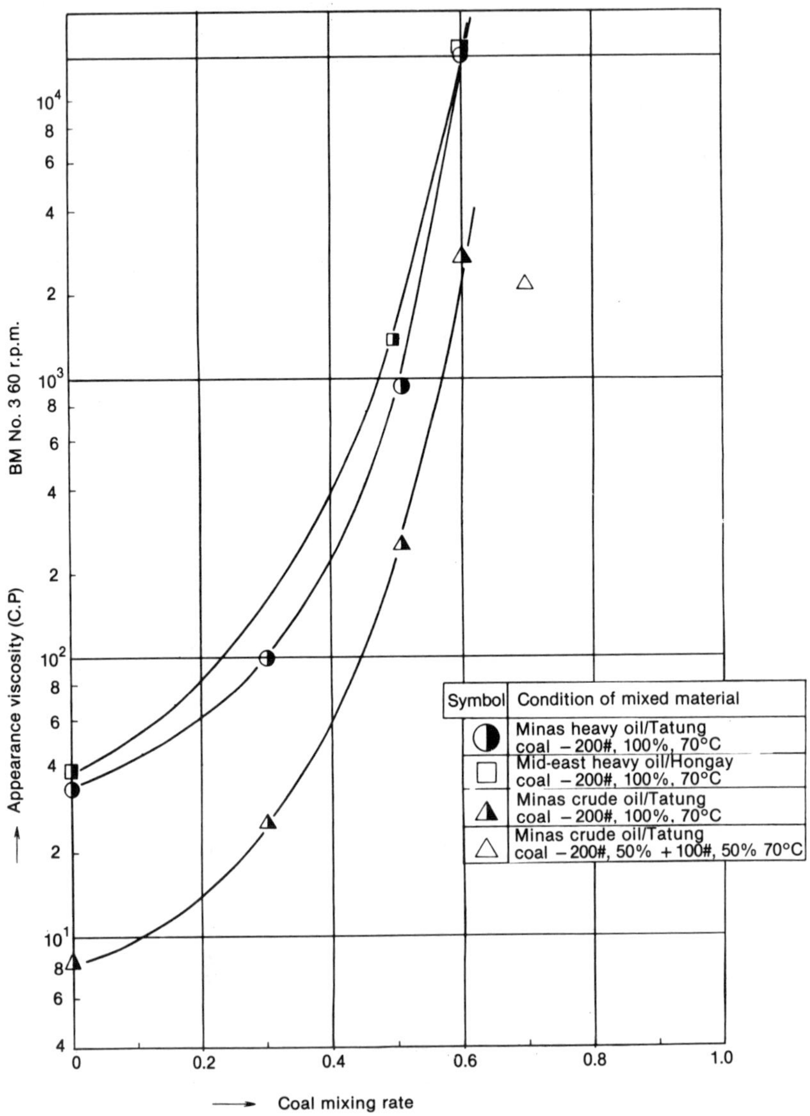

Figure 3-8. Influence of coal mixing rate on viscosity [24].

imized by using ceramic-coated modified gate valves, which are also "self-wiping" to reduce plugging problems. Viscosity of COM increases by as much as 10 times as the coal fraction is raised from 40% to 50%. The present system is designed for viscosities as high as 10,000 cp by increasing line diameter and minimizing line length. Plugging and settling

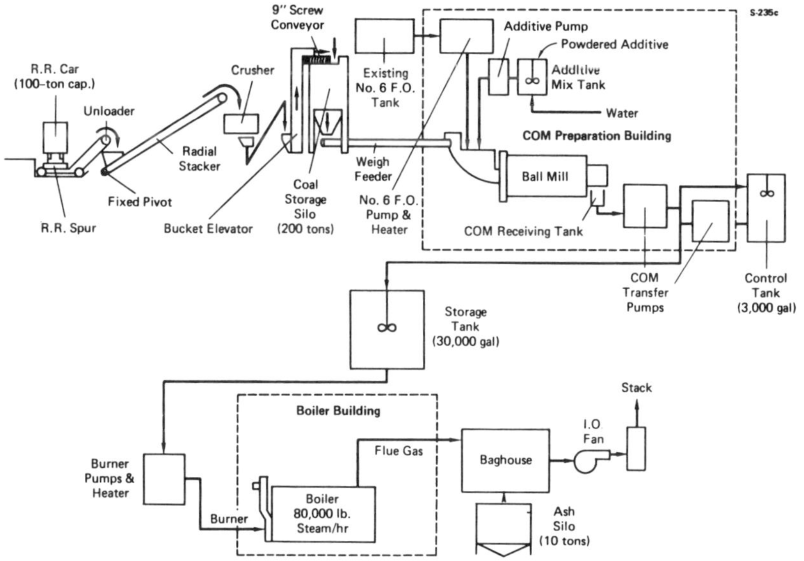

Figure 3-9. COM process flow diagram [47].

in valves, strainers, etc. have been addressed by eliminating potential flow stagnation regions using "self-wiping" seal designs and allowing for "duplex" cleanout capability so that production is maintained while strainers are cleaned. Specifications for the Accurex COM preparation plant are shown in Table 3-4. The COM will be fed to a COEN steam-atomized burner with flame-shaping capability for steering the flame away from the furnace walls.

BURNERS AND COMBUSTION

The DOE project is sponsoring tests of a wide variety of burner designs, along with special combustion tests to elucidate the details of the process of ejection of the atomized COM droplet from the nozzle, its ignition by the hot combustion chamber environment, and how variations in this process affect combustion efficiency and combustion products.

In one of the most detailed studies, Law et al. [48] at Northwestern University have been examining in microscopic scale the combustion properties of single droplets of COM. The COM jet spray is formed at a short distance downstream of the atomizing nozzle. When it moves to the combustor interior it entrains surrounding hot oxidizing air to promote vaporization and combustion of the fuel. The work of Law et al.

Table 3-4. COM Preparation Plant Specifications [47]

a. Feedstock Analysis

No. 6 Oil[a]		Coal[b]	
API Gravity	15.3	Moisture (% wt)	4.2
Sulfur (wt %)	2.22	Volatiles (% wt)	33.0
Flashpoint (PMCC °F)	204.0	Fixed Carbon (% wt)	54.0
Viscosity (SSF @ 122 °F)	247.0	Ash (% wt)	8.9
Pour point (F)	+50.0	Sulfur (% wt)	0.9
BS&W (% vol)	0.4	Ash Fusion (F° Temperature)	2700.0
Heating Value (Btu/lb)	17500.0	Hardgrove Grindability	68.0
		Heating Value (Btu/lb)	13368.0 As Received 13954.0 dry
		Size	1/2 × 0
		Origin: Clintwood seam, Conoway, Virginia	

b. 50% COM Analysis

Coal (wt %)	50.0
Oil (wt %)	46.25
Additive (wt %)	3.75
Sulfur (wt %)	1.5
Particle Size	80%–200 μ; 100%–48 μ
Heating Value (Btu/lb)	14,778

c. Selected Equipment Specifications

Preparation Plant	
Coal Storage Silo (tons)	200
Oil Storage Tank (gal)	300,000
COM Control Tank (gal)	3,000
COM Storage Tank (gal)	30,000
Ball Mill Production (ton/hr)	7.0
Motor (hp)	150
COM Transfer Pump-duplex	
Flowrate (gal/min)	25
Pressure (psig)	100
Speed (rpm)	84
Power (hp)	5
COM Tank Mixers	
Speed (rpm)	9.7
Power control tank (hp)	3
Power storage tank (hp)	30

[a]Supplied by Amerada-Hess Corporation.
[b]Supplied by Maryland Coal and Coke Company.

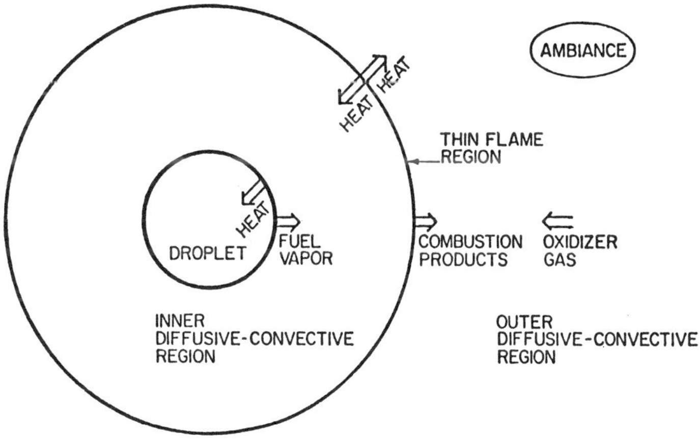

FLOW CONFIGURATION FOR SPHERICALLY SYMMETRIC DROPLET COMBUSTION

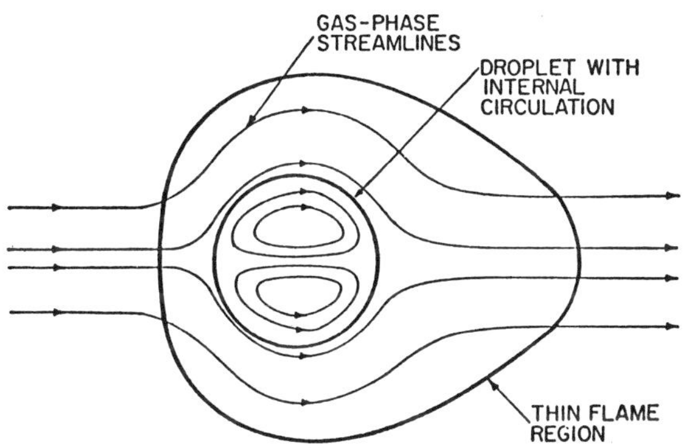

FLOW CONFIGURATION FOR DROPLET COMBUSTION IN THE PRESENCE OF FORCED CONVECTION

Figure 3-10. Combustion environment of COM droplet [48].

is concentrated on the individual droplet on the basis that these droplets are sufficiently distant from one another that they do not mutually interact and thereby affect each other's individual combustion. The suggested dynamics for the droplet are illustrated in Figure 3-10. When there is no relative motion between droplet and environment, stable combustion is maintained with the outwardly diffusing fuel vapor reacting in approximate stoichiometric ratio with the inwardly diffusing oxidizer gas at a thin enveloping flame front. When there is relative motion, there is some recirculatory motion within the droplet interior induced by the shearing stress acting at the droplet surface. Qualitatively different combustion characteristics may result depending on the intensity of the additional convective transport.

In their experiments, a COM droplet about 1 mm in diameter is suspended at the end of a quartz fiber and ignited through spark discharge. Subsequent combustion is studied through high-speed cinemicrophotography with framing rates around 1000 frames per second. After ignition there is a brief, steady burning period followed by intermittent ejection of small masses from the droplet. When most of the liquid fuel has been vaporized there is left behind an irregularly shaped agglomerate of coal particles adhering to the suspension fiber. There is apparently a mixed combustion with part of the coal particles dispersed and the rest agglomerated. These latter do not undergo complete combustion. Figure 3-11 shows the fractional amount of the coal particles agglomerated as a function of the initial droplet coal concentrations.

The work of Dooher et al. [25] at Adelphi University indicates that this problem may be relieved by introducing water into the COM. Photomicrographs of their coal-water-oil emulsions show fine coal particles adhering to the water droplets in a way similar to that of carbon black

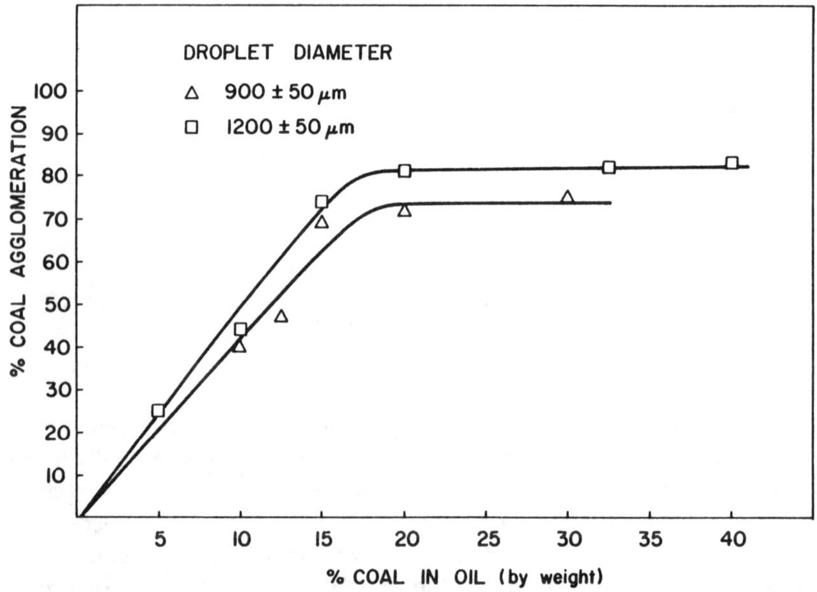

Figure 3-11. Fractional amount of coal particles agglomerated as a function of the initial content of coal [48].

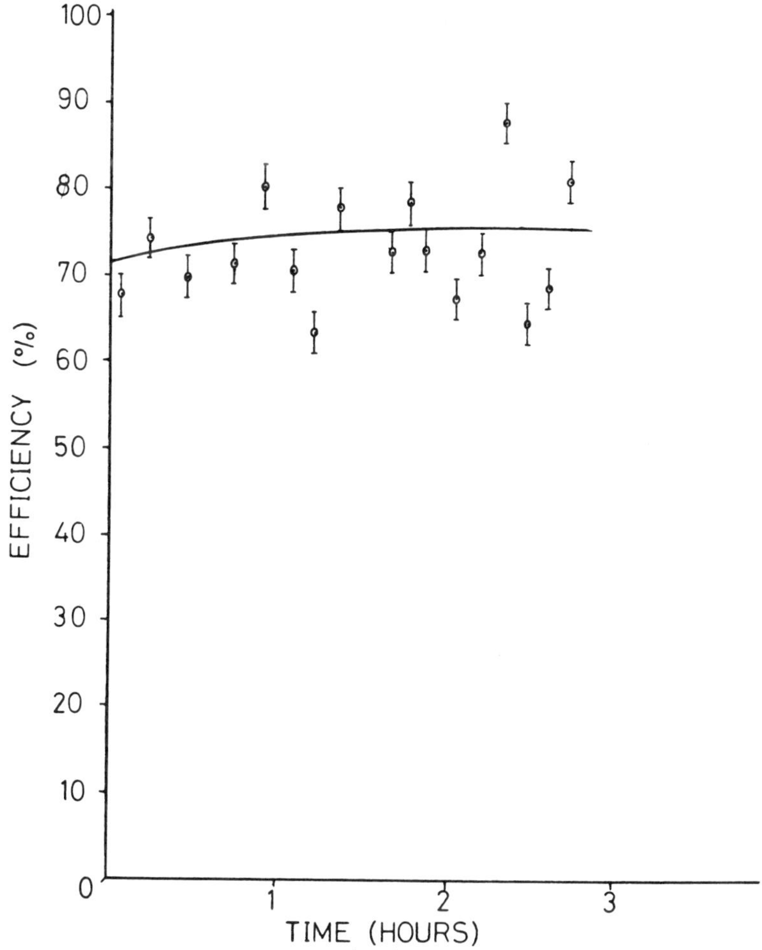

Figure 3-12. Combustion efficiency for No. 2 oil [25].

in water-oil emulsions. Large coal particles fill the interstices between the water droplets for added steric stabilization. Under combustion the water droplets give rise to miniexplosions absent for a simple coal-oil droplet. In the Adelphi experiments sample droplets were suspended on a thermocouple at the focus of a parabolic mirror with a high-intensity light source located at the focus of an opposing parabolic mirror. The effect of this intense irradiation was observed with a high-speed camera at 44,000 frames per second. When the light was focused on a coal-oil droplet only vaporization occurred. When the coal-water-oil droplet was heated in the same way there were many explosions as the water was vaporized, sending the coal particles off and completely rupturing the oil droplet.

To verify this conjecture of good combustion when water was incorporated, Dooher et al. [25] compared combustion efficiency in a heavily instrumented furnace using a Parr oxygen bomb in which the droplets are ignited. Figure 3-12 shows the oil combustion efficiency and Figure 3-13 shows the somewhat lower combustion efficiency for a 40-10-40 COM. However, the argument is not really complete without a similar test on a simple coal-oil mixture to test the prognosis implied by the re-

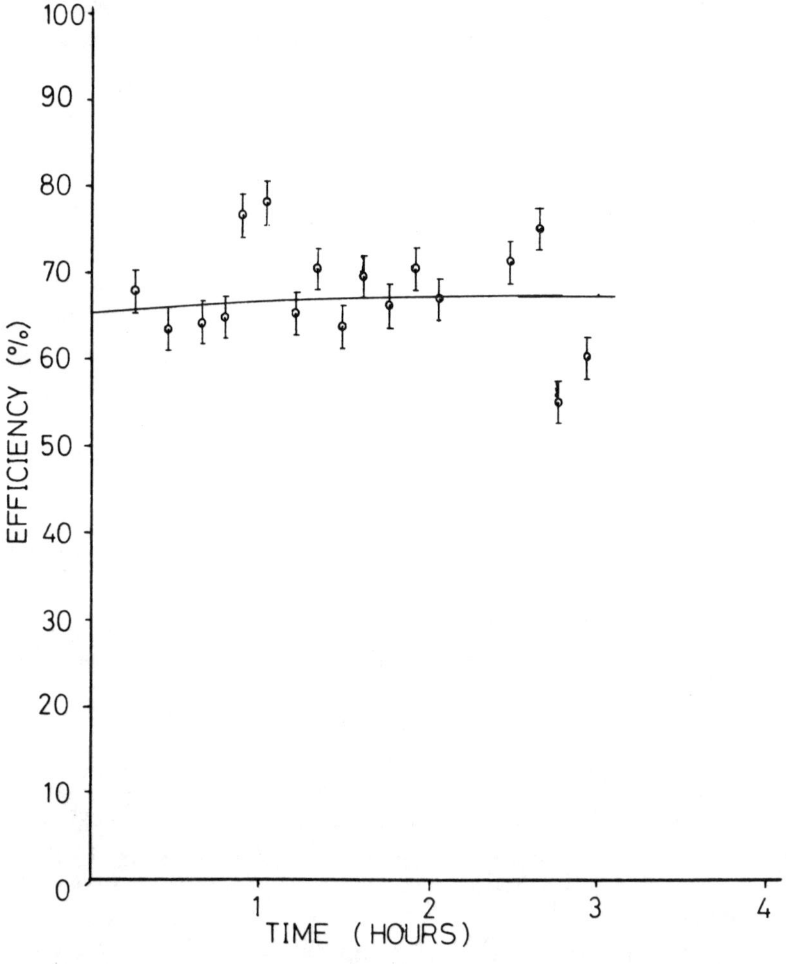

Figure 3-13. Combusion efficiency for Pennsylvania coal-water-oil emulsion at 44.5–11.1–44.5% [25].

sults of Law et al. [48]. One must also note that this prognosis of incomplete coal combustion is not consistent with the favorable testimony of the early COM work in the 1930s and 1940s.

On a more engineering basis, Demeter et al. [3] at PERC carried out combustion tests on 20% coal-in-oil using the burner head and nozzle shown in Figure 3-14. The fuel ignited readily and produced a very stable flame, which responded well to load changes. Excess air level was varied from 20–75% during the course of the 1000-hour test. Subsequent boiler efficiency tests, with results shown in Table 3-5, indicated that the boiler efficiency and combustion efficiency for COM were about the same as that obtained with fuel-oil alone. When firing the boiler at rated capacity (100 hp) with the standard nozzle and diffuser, the flames impinge on the refractory in the burner zone causing solid deposits to build in the combustion chamber. Increasing excess air level extended operating time, but eventually the boiler was shut down to remove chamber deposits. Some improvement was obtained by modifying the nozzle to provide a narrower flame pattern with reduced impingement. There was also some deterioration of the spray pattern due to the erosion of the relatively

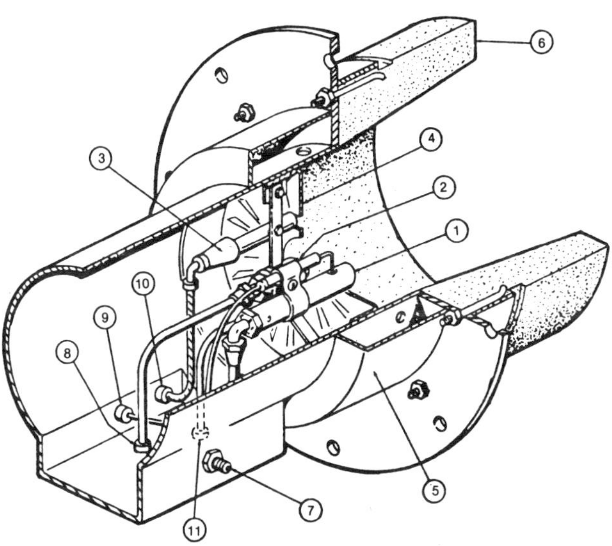

1. Gas Pilot and Ignition Electrode
2. Oil Nozzle
3. Flame Detector (Scanner)
4. Diffuser
5. Gas Manifold (Note Ports)
6. Refractory Oven
7. Pilot Gas Inlet
8. Atomizing Air Inlet
9. Electrode Lead
10. Scanner Lead
11. Fuel Oil Inlet

Cutaway showing details of the burner head.

Figure 3-14. COM combustion system for PERC firetube boiler.

Table 3-5. Comparison of No. 6 Fuel Oil and Coal-Oil Slurry Combustion Tests [3]

Fuel Type	No. 6 Oil	Slurry
Coal Concentration (wt %)	—	20
Coal size, <200 mesh (wt %)	—	95
Fuel Heating Value (Btu/lb)	18975	17904
Fuel Flowrate (lb/hr)	221	229
Fuel Pressure @ Burner (psig)	17	19
Atomizing Air Pressure (psig)	18	19
Feed Tank Temperature (fuel) (°F)	120	116
Fuel Temperature @ Burner (°F)	201	202
Stack Temperature (°F)	463	477
Stack Gas		
O_2 (%)	4.8	4.3
CO_2 (%)	11.4	11.5
CO (ppm)	50	81
NO_x (ppm)	223	273
SO_2 (ppm)	350	520
Steam Flowrate (lb/hr)	3570	3510
Excess Air (%)	27	24
Dust Loadings (lb/hr)	1.2	3.8
Carbon in Stack (%)	46.4	49.9
C Combustion Efficiency (%)	99.7	99.2
Boiler Efficiency (%)	80.1	79.9

small orifices in the brass burner nozzle. Use of a stainless steel nozzle in the latter part of the test gave a distinct improvement.

Combustion test results in the feasibility phase of the Accurex program were divided into three phases, described by Busch and Brown [46]. The burner used was an Accurex/IFRF wall-mounted unit capable of handling 1.5×10^6 Btu/hr. It is an experimental swirl block burner patterned after that of Beer [49]. For use with COM, two commercial fuel oil atomization nozzles were tried, as shown in Figure 3-15, the Delavan swirl-air (a) and the Sonic Development Corp. Sonicore (b).

In the first of three phases, the combustion stability of each fuel was evaluated and optimized. In all cases the nozzle position was 5 cm forward of the burner throat. However, atomization air pressure had to be optimized for each of the fuel mixtures. Poor atomization was caused by carbonaceous depositions in the water-cooled section of the burner. Required air atomization pressure ranged from 12.0 to 22.0 psig, depending on the fuel, usually 2–4 psig greater than the fuel delivery pressure. Secondary air was preheated to 150°C to promote solids ignition. Burner swirl optimized on parent fuels and COM was ⅝ for baseline coal and 0 for both COM and oil (no swirl). Earlier pilot tests at the

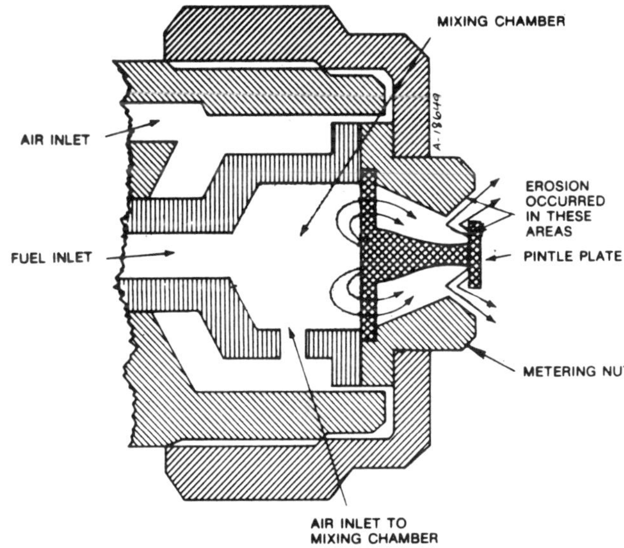

DELAVAN SWIRL-AIR NOZZLE, 33373-1. 60 gph, 70° SPRAY ANGLE, MILD STEEL CONSTRUCTION

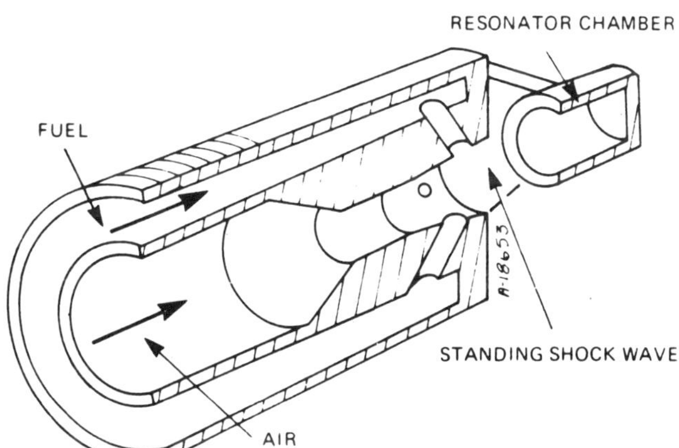

SONIC CORPORATION SONICORE NOZZLE, 281T-B-11 WITH STELLITE RESONATOR CHAMBER

Figure 3-15. Burner nozzles used by Accurex [46].

Table 3-6. COM Combustion Tests of Japan Electric Power Development Co. [24]

Item	Content of Research	Results
1. Clarification of fine COM atomizing conditions	Conduct COM atomizing tests by means of atomizer model. Take grain-size measurement of atomized COM to determine fine particles atomizing conditions. In view of COM viscosity, the atomizer is intermediate air-mixed steam atomizing type. Coal grain size is 200 mesh with pass rate of 70%.	(1) Atomized grain size of Minas crude oil lies in the range of 200-300μ, which shows that no reduction in efficiency of atomizing COM is witnessed. (2) Atomized grain size lies in the range of 500-600μ at injecting temperature of 70°C, against 200μ or less at higher temperature excepting two cases. (3) COM with water and additive provide ease in atomizing, generally. (4) COM with a high concentration of coal is not always poor in being atomized. Remarks: Correlations between (3) and (4) above are yet to be confirmed in the absence of further detailed test results.
2. Clarification of coal grain combustion speed	Grain size measurement to be taken of sample coal grains in an atmosphere made similar to that of an actual furnace. On the basis of the results thus obtained, estimate temperature within actual furnace, oxygen concentration and grain combustion time required. Test samples consist of either coal or coal immersed in heavy or crude oil. The latter two test samples are imitations of coal grains in COM. Coal has uniform grain size of 10–30 mg (3–4 mm) and is im-	(1) Time required for flame to die out almost parallels that required for volatile matter, oil content and inherent moisture of coal to disappear. (Namely, 85% of the initial amount of coal remains unburned during the said time.) (2) Time required for flame to die out does not depend greatly on oxygen concentration, but on temperature. (3) Flame of coal grains immersed in oil takes longer time to die out because of the oil content contained. (4) Speed of combustion of solid matter depends on temperature and oxygen concentration. (5) No significant differences are witnessed between the

	mersed in Minas crude or heavy oil for two days at normal pressure and at 70°C (inclusive of inherent moisture of coal).	combustion speed of coal grains immersed in oil and that of coal grains not immersed in oil.
3. Clogging of atomizer	Further study also in relation to test item (1).	Conclusion of experience gained on the atomizer test is summarized to the following three items (see Figure 6-1 for drawing of sample atomizer): (1) Clogging is witnessed in d_3 section in Figure 6-1. (2) Small atomizer develops frequent clogging. Clogging develops even during satisfactory atomizing. Causes are yet to be confirmed. (3) Large atomizer develops with clogging frequency of 15%. That is, about 15% of atomizing failed at the start of operation, due to clogging. However, no clogging was observed during operation once started. Therefore, the clogging seems to be attributable to insufficient purging of the atomizer after each use.

Accurex/EPA facility compared the nozzles. After 3 hours on 30% COM, the Delavan nozzle of 440 hardened stainless steel showed significant erosion. The same tests on the Stellite* Sonicore revealed no erosion. Hence, subsequent tests were all done with Sonicore of Figure 3-15b. The tests revealed significant differences among fuel types for combustion products that will be described in the next subsection. In the industrial-scale boiler demonstration program at Accurex, Shimizu and Derbidge [47] report that they will be trying steam atomization of the COM and a tungsten carbide tip for the burner nozzle.

At Japan's Electric Power Development Co., extensive combustion tests were undertaken and are described by Hatano in Table 3-6. Essentially they substantiate the beneficial effect of water in the COM for combustion. However, as yet they do not seem to have sufficient experience with the burner nozzles to report to the problems of erosion.

COM COMBUSTION PRODUCTS

There is serious concern over COM combustion products on two counts. First, there is the long-standing problem of coal emissions themselves, including particles and gas molecules that can abrade or corrode the surface of the heat exchanger tubes. Related to this problem there has been a continuing sequence of DOE programs designed to reveal and investigate new materials that can cope with a coal combustion environment. For example, there is a program just underway aimed at exploring ceramic coatings for gas and diesel turbines to be used with minimally processed coal-derived liquid fuels, of which high-coal-concentration COM would be a substantial example. Within the COM program combustion tests include analyses of the products for those that will cause these corrosion or abrasion problems. On a second count there is the problem of air pollution products in the stack effluent. These latter will be discussed in the section entitled Pollution Control.

In the PERC tests on a 100-hp firetube boiler, corrosion was investigated by direct examination of the interiors of test sections of the boiler taken from the second, third, fourth and fifth passes [3]. Metallographic analysis of the sections and chemical analysis of scrapings were used to determine how much abrasion and corrosion was experienced by these tubes. Weights and wet chemical analyses of the deposits are given in Table 3-7. Loose deposit removed by brushing contained a relatively

*Stellite—alloys of approximately 85% Co-15% Cr.

Table 3-7. Weights and Analyses of Deposits Removed from Boiler Tubes by Brushing, Abrading and Chemical Treatment [3]

Boiler Pass	2	3	4
Weight of Deposit Removed by Brushing ($mg/in.^2$)	38.2	91.0	61.0
Weight of Deposit Removed by Abrading ($mg/in.^2$)	38.6	12.1	27.6
Weight of Fe Removed Chemically ($mg/in.^2$)	7.0	3.2	14.6
Total	83.8	106.3	103.2
Analyses of Deposits			
(a) Removed by brushing (%)			
Al	5.3	5.1	2.2
C	10.7	12.2	15.4
Fe	31.9	26.6	33.5
Si	8.5	10.1	4.5
S	0.5	1.8	2.6
V	<0.1	<0.1	<0.1
Coal ash (calculated)	37.8	44.9	20.0
(b) Removed by abrading tool (%)			
Al	0.4	1.0	0.5
C	3.1	4.2	11.4
Fe	59.1	52.7	62.5
Si	0.6	1.1	0.9
S	0.6	1.0	1.1
V	<0.1	<0.1	<0.1
Coal ash (calculated)	2.7	4.9	4.0

large amount of carbon (10–15.4%) and, as determined by commercial analysis, ash contents of 20–45% on the various passes. Some X-ray analyses of the deposits in Table 3-8 show that the loose deposits were amorphous, with minor components being Fe_2O_3, Fe_3O_4, mullite and silica. Corrosion rates, calculated from the iron associated with this corrosion products, were between 0.23 and 0.40 mils per 1000 hour.

In the Florida Power and Dravo Corporation's working test of COM in producing power, Rodriguez and Sell [50] report the dependence of particulates and ash in the emissions on the coal concentration (Figure 3-16). It is quite clear that both particulates and ash are due almost entirely to the coal content of COM.

In the Accurex project, on 30% COM, Busch and Brown [46] took measurements of CO, CO_2, SO_2 and NO for their dependence on the varieties of parent fuels. However, they report only NO levels, as shown in Figure 3-17, for their dependence on percentage of excess air. There is clearly no substantial difference due to the oil base. The NO product appears related strictly to coal content and not ameliorated in COM except by the dilution to 30%. There is some distinction in NO as excess

Table 3-8. X-Ray Analyses of Deposits of Inner Surface (Fireside) of Boiler Tubes [3]

Location	Major Components	Minor Components
Pass No. 2		
(a) Loose surface layer	Amorphous	Mullite, quartz, Fe_2O_3, probably Fe_3O_4
(b) Layer near metal	Lepidocrocite FeO(OH) Goethite FeO(OH)	
Pass No. 3		
(a) Loose surface layer	Amorphous	Mullite, quartz, Fe_2O_3, probably Fe_3O_4
(b) Layer near metal	Goethite FeO(OH)	
Pass No. 4		
(a) Loose surface layer	Amorphous	Quartz, Fe_2O_3, goethite
(b) Layer near metal	Goethite	

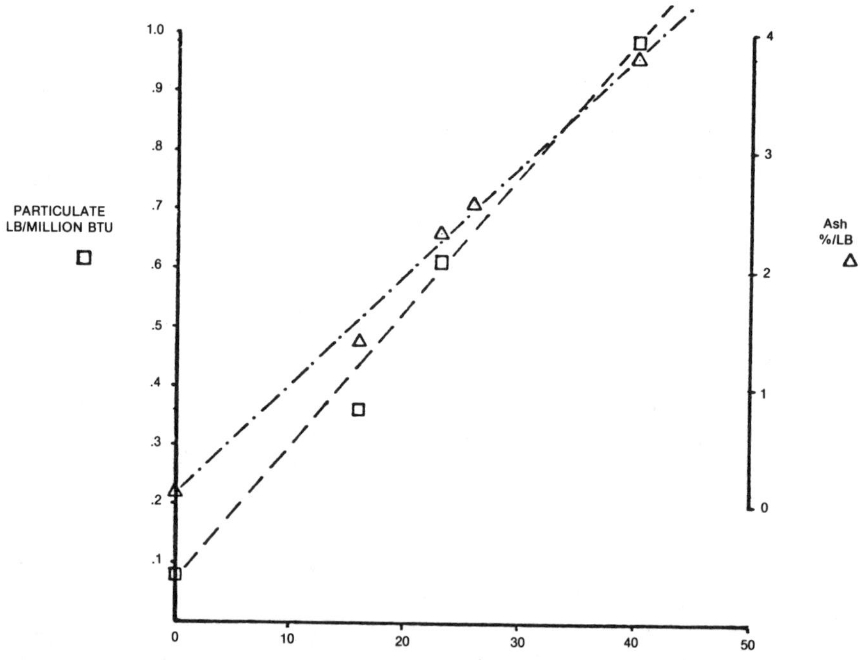

Figure 3-16. Dependence of particulate emissions on coal concentration in COM [50].

air is reduced. This dependence on air available for combustion may underlie a difference in NO for the two types of nozzle used in the burner shown in Figure 3-18.

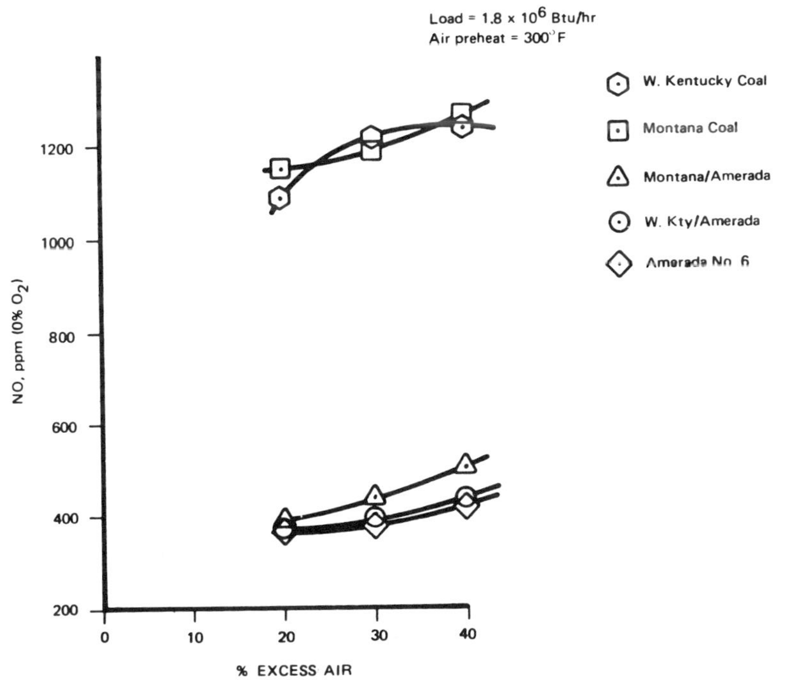

Figure 3-17. Emissions dependence on coal type and excess combustion air [46].

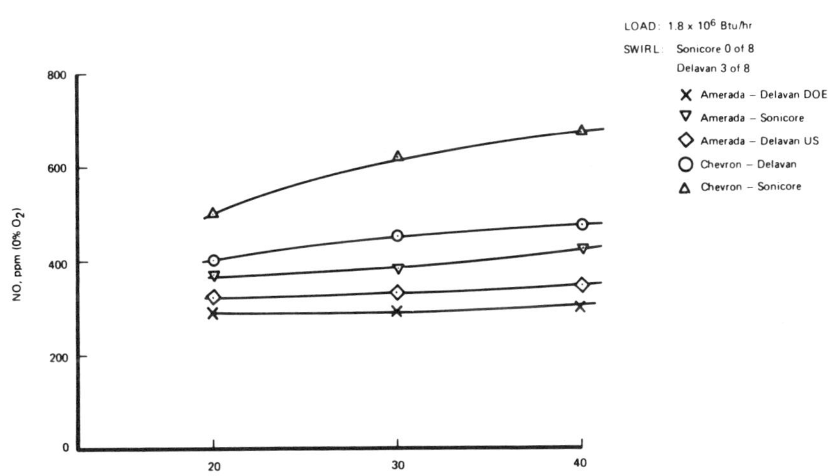

Figure 3-18. Nozzle influence on NO emissions [46].

HEAT EXCHANGE EQUIPMENT AND BOILER PROPERTIES

As yet there have been no new concepts for heat exchangers and other boiler equipment designed to either accommodate or take advantage of special COM properties. Its combustion properties being clearly a mixture of those for coal and oil, it is difficult to discern any special requirements other than those that could be met by materials capable of withstanding better the onslaught of the coal combustion products. Essentially, the COM-related boiler tests have all been carried out on existing boilers with no substantial modification of the basic heat exchange equipment.

The dependence of boiler operating efficiency on coal concentration at very low coal concentrations has been studied by Essenhigh et al. [51] in a hot wall furnace. As shown in Figure 3-19, the furnace has 40 ft^3 of combustion volume and fires at 1.25 million Btu/hr (approximately 0.35 MW). There are water-cooled tubes on the floor of the furnace to simulate the load of an industrial continuous thermal processing furnace. These tubes also were used to measure heat flux profiles and thermal efficiencies. There were two series of experiments, the first including 10% water in the mixture. In both series efficiency increased with coal percentage, reaching a peak at about 10% coal as shown in Figure 3-20. The water content definitely depressed results on this measure. As shown in Figure 3-21, the shape of the efficiency versus concentration results is reflected in heat flux and furnace temperatures, which vary with concentrations at one end of the furnace and oppositely at the other. Hence, it may be an artifact of the furnace design and, therefore, not generically extensible to all furnaces.

At PERC, on the 100-hp test firetube boiler, Ekmann et al. [52] used measurements of irradiance and first-pass temperatures to examine heat transfer patterns. Extensive data are summarized in Table 3-9 for oil, 20% COM and two runs of 30% COM. Stack temperature is adjusted to eliminate the dilution effect of the varying levels of excess air. Percent conversion of fuel nitrogen to NO_x assumes no atmospheric fixation. Nearly identical excess air was used for COM and oil tests. No degradation of performance was caused by air up to 30%. Stack temperatures were lower for COM than for air, indicating somewhat improved heat transfer since heat input and combustion performance were essentially equal. Levels of CO were low to indicate good COM combustion throughout the test sequence. Levels of SO_2 tracked the increasing sulfur content in COM. However, nitrogen oxide levels did not follow COM's

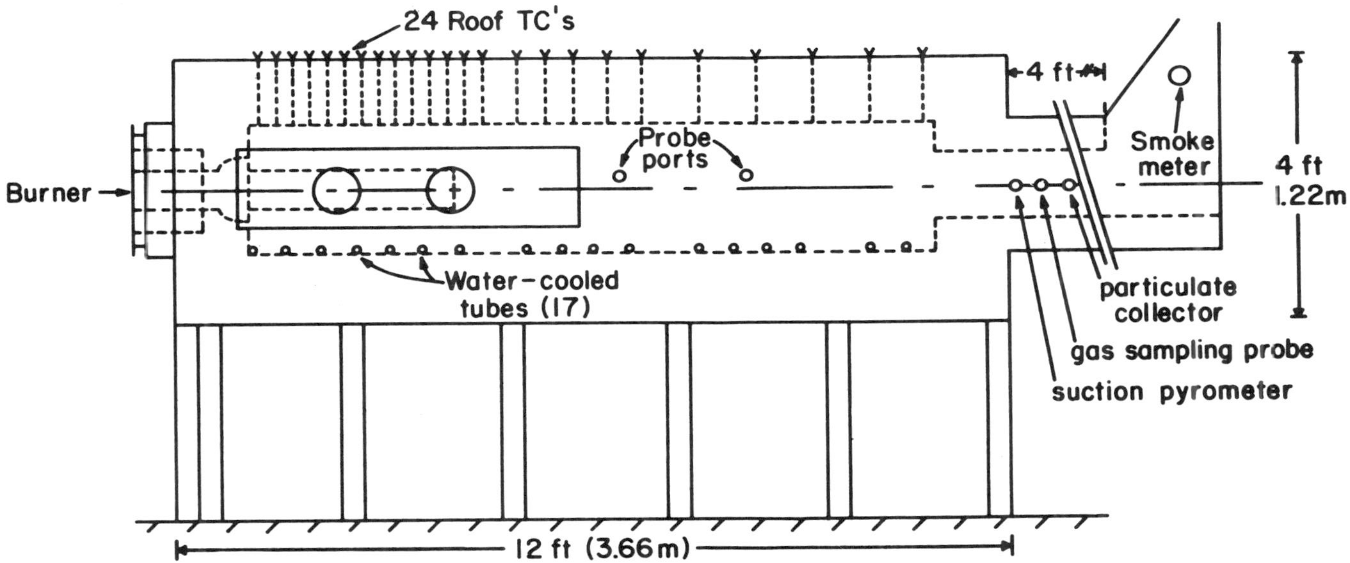

Figure 3-19. Schematic of furnace, illustrating external and internal (dotted line) dimensions and locations of monitor probes (gas analysis, temperature, solids and smoke) [51].

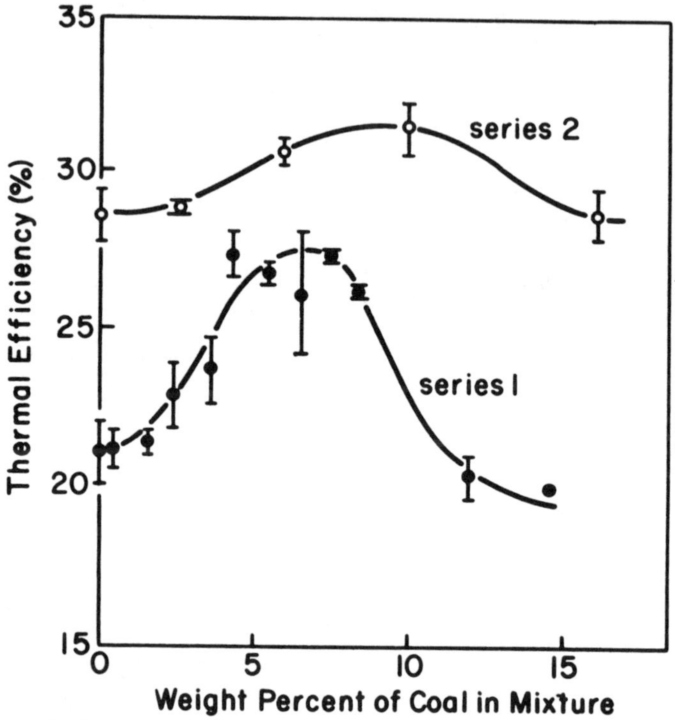

Figure 3-20. Variation of thermal efficiency with weight percent of coal in COM, for series 1 (with 10% water) and series 2 experiments. Bars denote standard error [51].

increase in fuel nitrogen shown in Table 3-9b. The flame diffuser's D_2 and D_3 reduced substantially the conversion to NO_x. Higher opacity readings for COM reflected increased ash content and also correlated with mass emission rates and carbon content of the collected ash.

Calculated values of efficiencies compacted by the heat loss method for eight oil and eight COM tests are listed in Table 3-10a. Analyses of the fuels used are in Table 3-10b. In all but one case the efficiency value for COM exceeds that for oil in the comparable test (see Table 3-9 for test conditions). Much of this improvement may well be attributed to the higher radiant fluxes of the COM flames, as listed in Table 3-9. Flame emissions measured in various ways show an increase with coal content that would indicate an increase in combustion chamber radiant heat transfer.

In the relatively short working test by the Florida Power Corporation, the test unit number 1 is a 2.5 million lb/hr, tangentially fired Combustion Engineering boiler (Figure 3-22). Its superheat/reheat design tem-

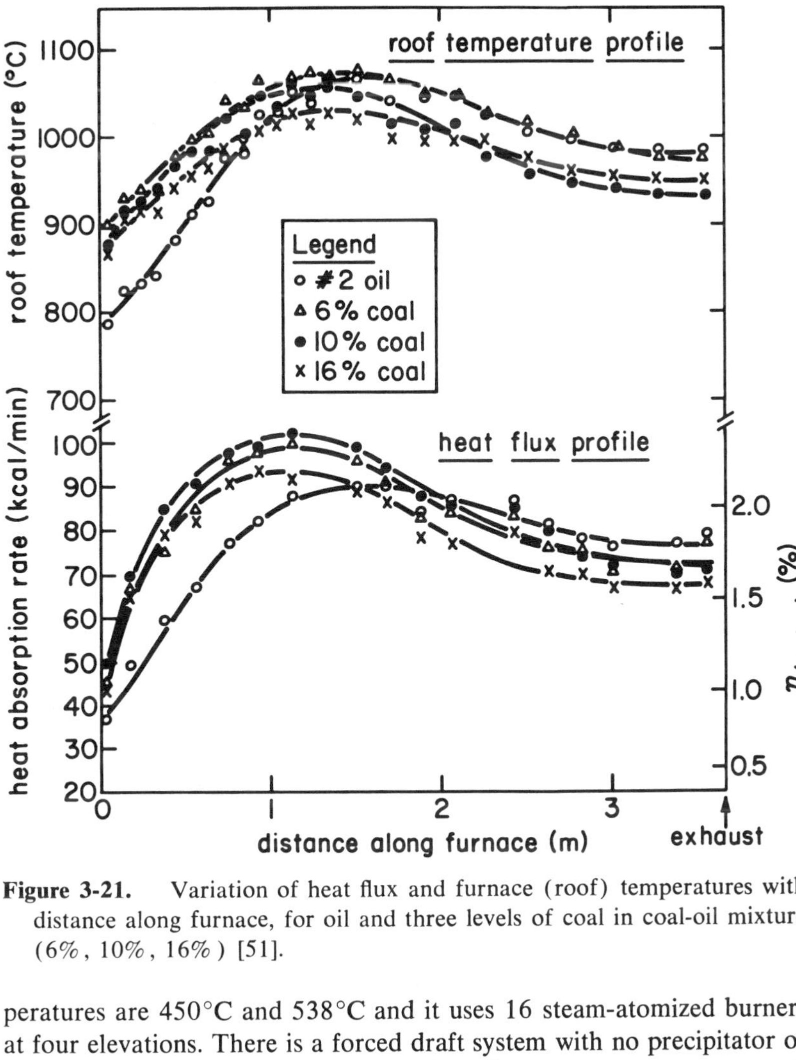

Figure 3-21. Variation of heat flux and furnace (roof) temperatures with distance along furnace, for oil and three levels of coal in coal-oil mixture (6%, 10%, 16%) [51].

peratures are 450°C and 538°C and it uses 16 steam-atomized burners at four elevations. There is a forced draft system with no precipitator or other pollution control equipment. Boiler exit gas temperatures approach 140°C at full load for 100% oil firing. Burners are manifolded so that excess fuel is returned back to the fuel handling system. Header pressure is maintained at 300 psig with additional burner pressure of 150 psig. Atomizing steam is kept at a 25-psig differential. Megawatt output capability for the different COM coal concentration is presented in Table 3-11 with an analysis of the fuels used in Table 3-11b. In Figure 3-23 the power output is compared with a theoretical curve corresponding to the maintenance of a constant fuel flow.

Table 3-9a. Boiler Operational Data [52]

Run No.	Fuel	Test Conditions Diffuser/% O_2/AA[a]/FR[b]	Percent Excess Air	Stack Temperature (adjusted to 0% excess air) (°F)	Opacity (%)	Irradiance (MW/cm^2-Sr)	Two-Color Temperature (maximum) (°F)
S-O-23	No. 6	D2/3.0%/15/15	11.0	499	2.2	79.2	2620
S-O-24	No. 6	D1/3.0%/30/15	8.5	475	3.4	88.5	2850
S-O-25	No. 6	D2/5.5%/30/15	24.0	566	—	45.8	2930
S-O-26	No. 6	D1/5.5%/15/15	24.0	607	—	93.5	2950
S-O-27	No. 6	D2/3.0%/30/27	8.5	694	7.3	80.5	2850
S-O-28	No. 6	D1/3.0%/15/27	7.0	611	3.0	52.0	3000
S-O-29	No. 6	D2/5.5%/15/27	24.0	671	4.0	47.5	3000
S-O-30	No. 6	D1/5.5%/30/27	24.0	701	3.2	71.5	2900
S-O/C-31	20% COM	D3/3.0%/15/15	13.0	482	10.4	59.2	2520
S-O/C-32	20% COM	D2/3.0%/30/15	10.0	478	11.9	220.0	2790
S-O/C-33	20% COM	D3/5.5%/30/15	26.0	572	9.2	216.8	2880
S-O/C-34	20% COM	D2/5.5%/15/15	29.0	530	7.1	171.5	2880
S-O/C-35	20% COM	D3/3.0%/30/27	14.0	591	18.0	201.2	3050
S-O/C-36	20% COM	D2/3.0%/15/27	12.5	522	8.0	120.0	3350
S-O/C-37	20% COM	D3/5.5%/15/27	23.0	545	13.8	116.2	3240
S-O/C-38	20% COM	D2/5.5%/30/27	26.0	599	12.8	124.8	2905
S-O/C-39	30% COM	D3/5.5%/30/27	27.0	629	20.2	—	—
S-O/C-40	30% COM	D3/3.0%/30/27	16.0	532	18.0	—	—

Table 3-9b. Boiler Operational Data [52]

Run No.	Fuel	Test Conditions Diffuser/% O_2/ AA[a]/FR[b]	Exhaust Gas Conc. (lb/MBtu) NO_x	CO	SO_2	% Conversion of Fuel N to NO_x
S-O-23	No. 6	D2/3.0%/15/15	0.221	0.090	0.57	61.6
S-O-24	No. 6	D1/3.0%/30/15	0.175	0.077	0.60	46.8
S-O-25	No. 6	D2/5.5%/30/15	0.263	0.050	0.60	69.0
S-O-26	No. 6	D1/5.5%/15/15	0.282	0.023	0.59	75.2
S-O-27	No. 6	D2/3.0%/30/27	0.225	0.270	0.60	59.4
S-O-28	No. 6	D1/3.0%/15/27	0.324	0.368	0.59	86.1
S-O-29	No. 6	D2/5.5%/15/27	0.301	0.074	0.62	77.6
S-O-30	No. 6	D1/5.5%/30/27	0.282	0.052	0.62	72.7
S-O/C-31	20% COM	D3/3.0%/15/15	0.174	0.303	0.85	21.9
S-O/C-32	20% COM	D2/3.0%/30/15	0.326	0.029	0.86	40.9
S-O/C-33	20% COM	D3/5.5%/30/15	0.188	0.085	0.86	23.3
S-O/C-34	20% COM	D2/5.5%/15/15	0.401	0.043	0.86	50.4
S-O/C-35	20% COM	D3/3.0%/30/27	0.331	0.234	0.85	41.6
S-O/C-36	20% COM	D2/3.0%/15/27	0.294	0.076	0.84	—
S-O/C-37	20% COM	D3/5.5%/15/27	0.288	0.062	0.85	36.4
S-O/C-38	20% COM	D2/5.5%/30/27	0.354	0.045	0.87	43.5
S-O/C-39	30% COM	D3/5.5%/30/27	0.375	—	1.03	38.8
S-O/C-40	30% COM	D3/3.0%/30/27	0.356	—	1.02	29.3

[a]Atomizing air pressure (psig).
[b]Firing rate.

The departure of measured results was expected due to the lower heating value of the coal-oil mixture relative to oil and the operating restriction on fuel flow imposed for comparison with oil. Increasing the orifice size on the burner can compensate for lower COM heating value to maintain the rated load of the unit. There were no problems involved with burning the COM. The fuel was ignited with a warm boiler to determine whether the ignitors could start combustion. The boiler was brought up to light load with No. 2 oil, then brought up to a maximum load with the mixture. Boiler response time was 15–20 minutes for fuel switching at various loads and coal concentrations. Maximum load for fuel switching was 315 MW with no observed problems. Burner combustion seemed to be stable and appeared to show high efficiency. The COM flame was visibly more radiant than for 100% oil, confirming the observations of Ekmann et al [52].

POLLUTION CONTROL

When COM emissions appear in the stack they often become a concern with regard to the normal U.S. Environmental Protection Agency

Table 3-10. Fuel Combustion Efficiency [52]

Weight Percent	Fuel			
	No. 6 Oil	20% COM	30% COM	Coal
C	86.2	84.0	83.1	75.3
H	11.9	10.8	10.1	5.3
N	0.23	0.4	0.6	1.5
S	0.60	0.88	1.0	2.0
Ash	0.13	1.3	2.1	7.1
0 (dif)	0.94	2.62	3.1	8.9
Heating Value (Btu/lb)	19,060	18,120.0	17,430	13,460

Carbon Conversion and Boiler Efficiency

Test	Fuel	Carbon Conversion (%)	Efficiency (heat loss method) (%)
S-O-23	No. 6 Oil	N/A[a]	83.0
S-O-24	No. 6 Oil	N/A	82.3
S-O-25	No. 6 Oil	N/A	80.7
S-O-26	No. 6 Oil	N/A	80.3
S-O-27	No. 6 Oil	99.4	78.9
S-O-28	No. 6 Oil	99.9	81.7
S-O-29	No. 6 Oil	99.9	79.5
S-O-30	No. 6 Oil	N/A	78.9
S-O/C-31	20% COM	N/A	83.7
S-O/C-32	20% COM	99.8	83.5
S-O/C-33	20% COM	99.6	80.4
S-O/C-34	20% COM	99.7	80.8
S-O/C-35	20% COM	N/A	81.3
S-O/C-36	20% COM	99.3	82.8
S-O/C-37	20% COM	98.5	81.1
S-O/C-38	20% COM	99.5	80.2
S-O/C-39	30% COM	98.5	79.8
S-O/C-40	30% COM	N/A	83.0

[a]Not available; carbon in ash for efficiency calculation estimated from similar test.

Figure 3-22. Florida Power Crystal River unit #1, tangential firing [50].

(EPA) standards governing utility boiler operations. As pointed out by Freedman et al. [18] in their description of DOE's COM development program, it is expected that COM fuels will have characteristics that are the weighted averages of the concentrations in constituent fuels for most properties. The COM will have ash and sulfur contents between those of the coal and oil of which it is composed. For example, Figure 3-24 estimates and compares to the EPA standard the SO_2 emissions as a function of coal concentration in a COM made from a 0.5%-by-weight sulfur oil and coals of different sulfur content between 1% and 3%. At the

Table 3-11. Power Generation Characteristics of COM [50]

(a) Megawatt Output Versus Percentage Coal

% Coal	Btu/lb	MW
0	18,500	383
23	17,200	375
26	16,000	350
40	16,200	315
45	—	260

NOTE: These are not boiler derating values. Many parameters were changed during each one of the tests. Of major significance were varying fuel flow and heating values of oil and coal in the mix. This is evident in the Btu/lb column. For further clarification see Figure 3-15.

(b) Typical FPC Coal and Oil Analyses

	Coal	Low-Sulfur Coal	No. 6 Oil
Hydrogen (%)	5.59	4.68	12.2
Carbon	64.53	71.89	81.3
Nitrogen	1.39	1.58	0.5
Oxygen	17.48	6.77	4.0
Sulfur	2.28	1.28	2.0
Sodium			0.004
Vanadium			0.016
Moisture	10.43	10.10	(0.157)
Ash	8.73	3.70	0.211
Btu/lb	11523	12723	18911
Viscosity SUS @ 150°F			734
210°F			209
300°F			77
API Gravity @ 60°F			11.1
Specific Gravity @ 60°F			.992
% Iron in Ash	19.01	24.40	
Ash Fusion			
Deformation (°F)	2100	2340	
Softening (°F)	2320	2390	
Hemispherical (°F)	2365	2430	
Fluid (°F)	2420	2475	

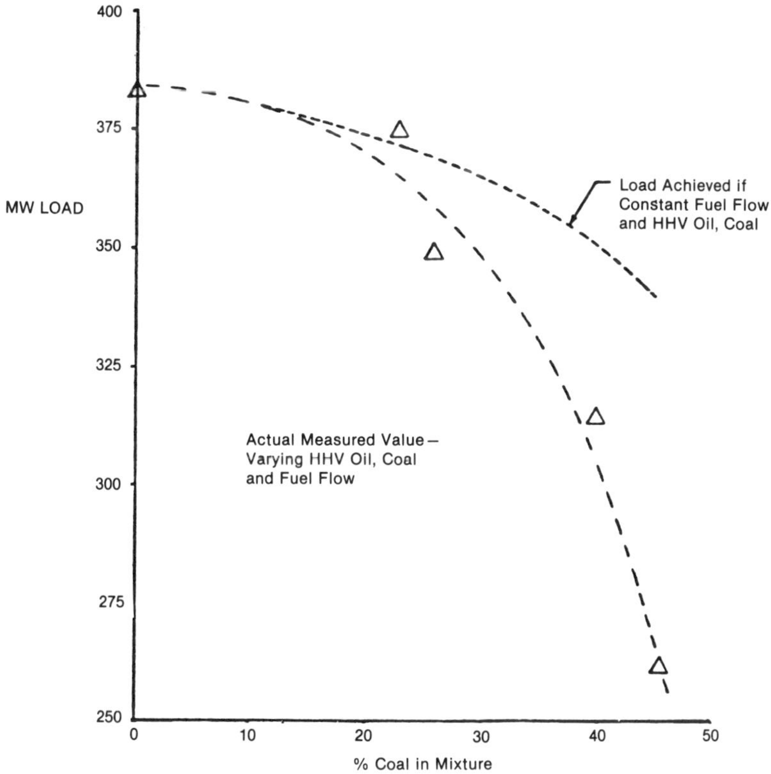

Figure 3-23. Megawatt load versus percentage coal in COM [50].

50% coal concentration one still may have to reduce SO_2 emissions by as much as 50–60%.

For the PERC test on a 100-hp firetube boiler, Ekmann et al. [52] observed stack emissions explicitly for opacity, which is correlated with fly ash emission, carbon content and particle size distribution. These results are shown in Table 3-12. The relationship of opacity to mass emission rates was established on each source by collecting an isokinetic stack sample. The mass to operating relationship is valid as long as the particle size of the fly ash is held constant. Using calibration points and apparent statistical relationship of opacity to dust loading, a standard error estimate of 65 mg/m^3 was calculated. The results in Table 3-12, although their data points were scattered, clearly indicate higher levels for COM than for oil.

In the Florida Power Company working test, Rodriguez and Sell [50] concluded that to control particulate emission, a precipitator, suitably scaled to the coal concentration, was plainly necessary. Their Crystal

Table 3-12. Opacity, Dust Loading and Carbon Content in Stack [52]

Run	Opacity (%)	Grain/ft³	lb/hr	C (%)	Stack Volume (cfm)
S-O-23	2.2	0.036	0.246	—	791
S-O-24	3.4	0.047	0.313	—	786
S-O-25	—	0.018	0.140	—	908
S-O-28	3.0	0.039	0.530	39.8	1569
S-O-29	4.0	0.028	0.434	55.1	1764
S-O-30	3.2	0.026	0.410	—	1835
S-O/C-31	10.4	—	—	87.9	—
S-O/C-32	11.9	0.173	1.11	17.7	747
S-O/C-33	9.2	0.151	1.12	37.4	863
S-O/C-34	7.1	0.144	1.06	30.2	861
S-O/C-35	18	0.188	2.48	42.1	1532
S-O/C-36	8.0	—	—	29.4	—
S-O/C-37	13.8	0.450	5.37	46.5	1403
S-O/C-38	12.8	0.174	2.36	—	1580
S-O/C-39	18	0.5	7.05	49.84	1642
S-O/C-40	19.5	0.28	2.02	—	842
S-O/C-41	10	0.137	0.987	40.1	842
S-O/C-42	10	—	1.72	57.5	—

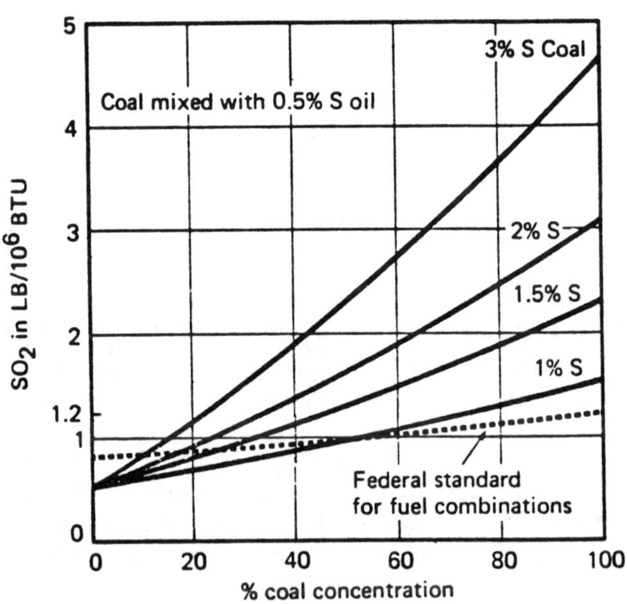

Figure 3-24. Potential sulfur dioxide emission from COM combustion [18].

River power plant complex, including the test unit No. 1, is surrounded by four monitoring stations, as shown in Figure 3-25. For each firing test a 24-hour sample was obtained providing two types of data:

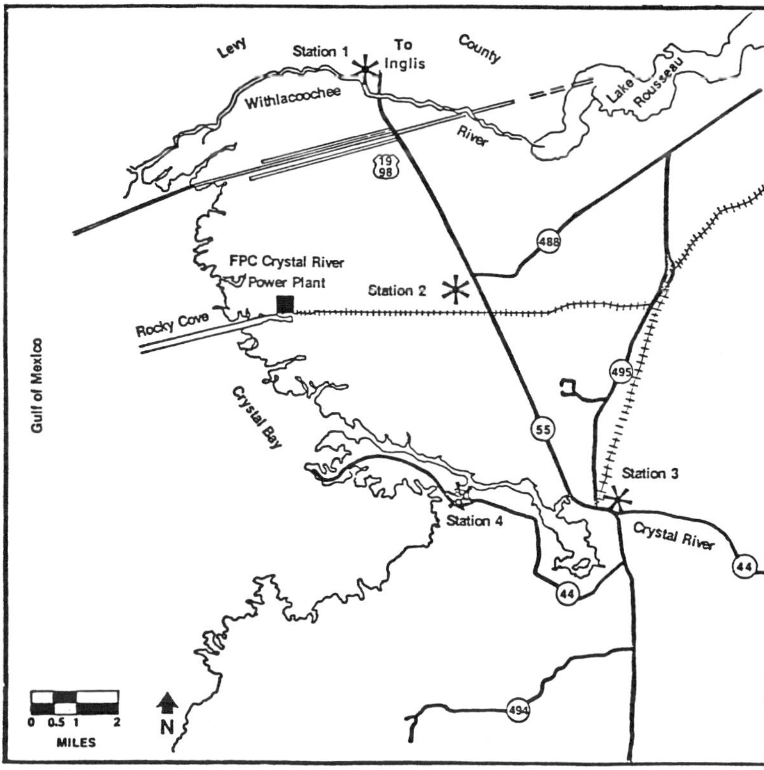

Figure 3-25. Crystal River power plant complex ambient air monitoring stations [50].

1. total suspended particulate matter, and
2. maximum sulfur dioxide concentration.

In Figure 3-26 the data on particulate matter are shown as averages for ambient monitoring stations 1, 2 and 4. Data for station 3 were obscured by consistently high background values due to a major highway and, therefore, were excluded. As shown, the ash and particulate matter are roughly proportional to the percentage of coal in the mixture. Hence, precipitator sizing could be determined based on the maximum percentage of coal in the slurry. Rather detailed results on particulate emissions were obtained in the General Motors study [20] for a long period of operation at COM coal concentration levels fluctuating between 40 and 50%. These results, spanning levels of 20–200 $\mu g/m^3$, as illustrated in Figure 3-27, indicate the wide variability possible and also the consistency of the Florida Power-Dravo results. They do, however, cast doubt

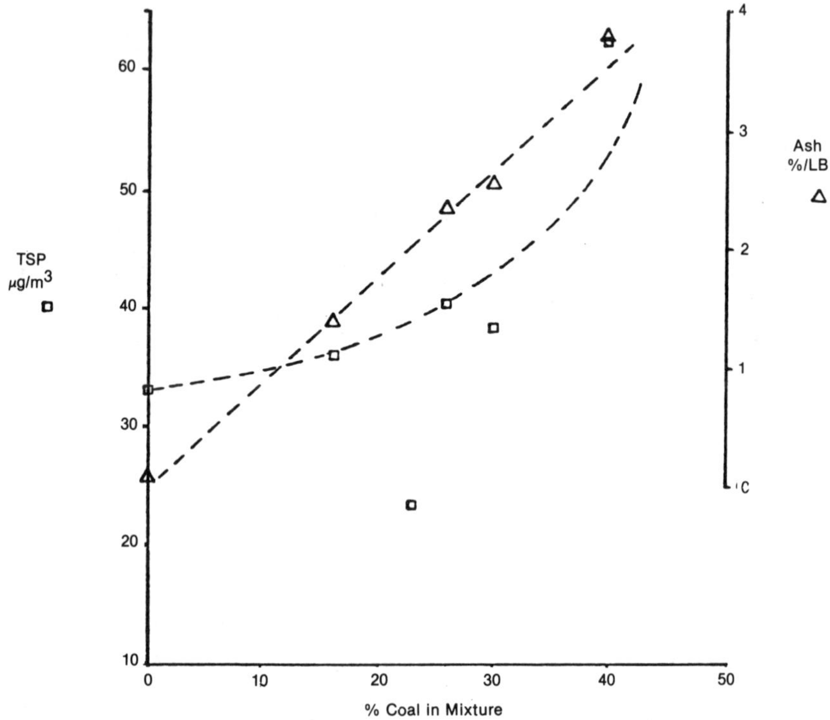

Figure 3-26. Total suspended particulate (TSP) matter in ambient vs percentage coal in COM [50].

on the possibility of inferring a quantitative measure of the dependence of particulate emissions on coal concentration in COM.

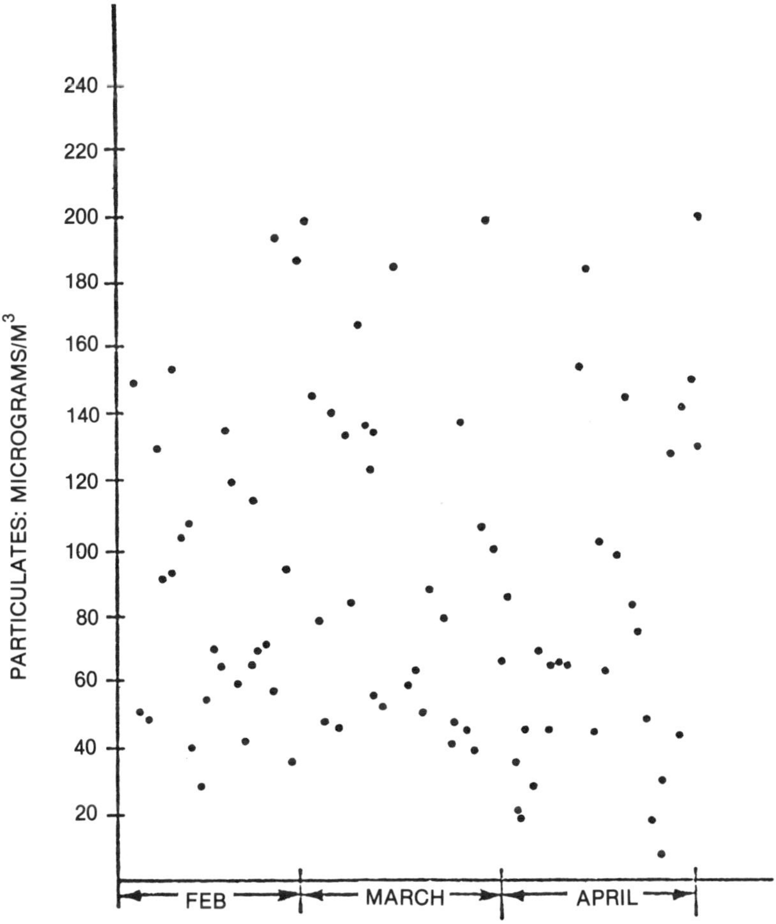

Figure 3-27. Particulate emission in General Motors tests at 40% COM levels.

CHAPTER 4

ECONOMIC CONSIDERATIONS

In the spring of 1978, as related by Ehrlich [53], the Electric Power Research Institute concluded that *in some special cases* coal-oil mixtures were an answer to the electric utility's questions about how to utilize $100 billion worth of units not originally designed to burn coal. The following conditions were attached:

1. The boiler was really a coal design (large furnace and wide convection tube spacing).
2. Coal handling is either impossible or impractical.
3. Regulators will permit coal-oil burning.
4. An infrastructure can be developed.

By February of 1979, the magazine *Power* considered that neither utilities nor industry had much motivation to burn COM unless oil consumption was restricted by the government and exports of oil were reduced by the major exporting nations [54]. Such views primarily reflect the still marginal economics for the difference between burning oil and burning COM, when the investment and operating costs for handling coal are considered.

The prospects of switching to coal are generally regarded by utilities and industries as decidedly unpleasant. As described by Bergman et al. [55], for example, the derating of thermal processing equipment, disruption in plant layout and land use patterns, costly delays, difficulties in ensuring stable coal supplies, and intractable environmental problems would accompany the introduction of direct coal firing. They considered that COM offered an intermediate strategy for carrying out a smooth transition from oil-firing to coal-firing but cited the possibility of preparing COM offsite as a further amelioration.

Described first in this section are exemplary sets of costs associated with the primary COM model being considered now, that is, individual onsite COM preparation. This, of course, imposes the major costs to be balanced against the oil-coal price differential. Secondly, the value of switching to COM as a utility investment is explored according to some analyses based on the available cost data and reasonable speculations about future fuel costs.

COSTS OF SWITCHING TO COM FROM OIL

Some estimates of costs for switching to COM were made by Eberle et al. [56,57] for a wide range of power plants grouped according to the following categories.

- boilers designed for coal and currently using oil
- industrial boilers producing saturated steam or small amounts of superheat
- utility boiler designed to fire oil
- gas-fired utility boiler where the majority are designed for "limited oil firing" or "future coal"
- gas-fired industrial boiler typically designed for extremely high liberation rates (30,000–60,000 Btu/hr/ft^3)

The capital cost of modifications and the COM preparation plant were estimated for various sizes. Table 4-1 shows estimated costs of industrial and utility steam generators of various sizes, including required heating value of fuel and the steam flow. Among the modifications listed, the precipitator usually contributed the largest proportion of costs, according to Hickman [58] of Forney Engineering. There is great cost variability for these items since all retrofit equipment depends heavily on the specific original system. Typically, a precipitator for a 100-MW unit would run a $1 million (1976) before installation, which itself could add another $1 million, for a total of 25% of the cost. Burners and pumping equipment would be next, followed by fly ash-handling equipment. Table 4-2 shows the estimated capital cost of various-sized COM preparation plants based on the following general relations:

1. Coal *unloading* facility for a small prep plant is estimated at $600,000.
2. Coal handling and storage facility:
   $400,000 for small prep plants
   $19,000/ton/hr of coal for larger prep plants
3. Oil and COM storage are estimated at $6.30/barrel.
4. Pulverizing and mixing:
   For 16 ton/hr of COM and less $1,000,000

Table 4-1. Capital Cost ($/kW) of Modifying a Steam Generator to Burn COM [56,57]

Unit Size (MW)	5[4]	10[4]	20[4]	30[4]	40[4]	80[4]	100	200	300	400	500	600	700	800
Fuel Required (MBtu/hr)	50	100	200	300	400	800	1,000	2,000	3,000	4,000	5,000	6,000	7,000	8,000
Steam Flow (lb/hr)	40,000	80,000	160,000	240,000	320,000	640,000	800,000							
Capital Cost (5/kw)														
1. Designed for coal-burning oil[1]	12	15	18	20	25	28	23	15	12	10	9	7	6	5.5
2. Designed for oil[2]	51	56	61	68	76	90	80	62	52	46	40	35	31	22
3. Designed for gas[3]	55	60	65	72	80	94	84	66	56	50	44	39	35	30

[1]Includes adding the following modifications:
1. Burner equipment
2. Burner safety system
3. Pumps, piping, etc.
4. Storage facilities

[2]Include adding the following modifications:
1. Precipitation
2. Soot blowers
3. Ash handling facilities
4. Burner equipment
5. Burner safety system
6. Pumps, piping, etc.

[3]Includes adding the following modifications:
1. Precipitation
2. Soot blowers
3. Ash handling facilities
4. Burner equipment
5. Burner safety system
6. Pumps, piping, etc.
7. Storage facilities

[4]These steam generators are package type units used in industry.

Table 4-2. Capital Cost ($) for a COM Preparation Plant [56,57]

Unit Size (MMBtu/hr)	50	100	200	300	400	800	1000	2000	3000	4000	5000	6000	7000
Coal Unloading $\times 10^6$	0.600	0.600	0.600	0.600	0.600	0.600	—	—	—	—	—	—	—
Coal Handling and Storage $\times 10^6$	0.400	0.400	0.400	0.400	0.400	0.400	0.400	0.745	1.117	1.490	1.862	2.234	2.607
Oil and COM Storage $\times 10^6$	0.045	0.089	0.178	0.267	0.356	0.712	0.892	1.784	2.676	3.568	4.460	5.352	6.244
Pulverizer and Mixing System $\times 10^6$	1.000	1.000	1.000	1.000	1.000	2.940	3.675	7.350	11.025	15.000	18.375	22.050	25.725
Total ($) $\times 10^6$	2.045	2.089	2.178	2.267	2.356	4.652	4.967	9.879	14.818	20.758	24.697	29.636	34.576

Up to 80 ton/hr $93,750/ton/hr
Larger plants $81,250 ton/hr

The operating and maintenance cost for the prep plant is estimated on the following basis:

1. Required power for handling coal is 10 kWh/ton of coal.
2. Required power for pulverizing and mixing is 30 kWh/ton of coal (80%-200 mesh).
3. For an industrial steam generator, coal-handling manpower required is two to five men per shift. For a utility-type system, it is eight men per day.
4. For industrial plants, the prep plant will be operated by the same men. Larger prep plants of utilities would need an additional five men per day.

Comparing estimates in Tables 4-1 and 4-2 we see that at a unit size of 7000 MMBtu/hr the capital costs for the COM preparation plant outweigh somewhat those for modifying the steam generator. For example, for an oil- and gas-burning boiler, the total new capital costs would be $70–80/do kW.

INVESTMENT VALUE FOR COM

Despite the seemingly high cost for the boiler modification and COM preparation, there is nevertheless a fair investment value based on the following parameters of an economic evaluation used by Eberle et al. [56,57]:

Year of Operation	15
Capacity Factor	80%
Derating Factor	Gas = 0.75 oil = 0.9 coal = 1.0
Natural Gas Cost	$2.0/MMBtu
Natural Gas Escalation	6%/yr
Oil Cost	$2.0/MMBtu
Oil Escalation	10%/yr
Coal Cost	$1.0/MMBtu
Coal Escalation	6%yr
Coal Concentration	50%
Water Concentration	6.5%
Oil Heating Value	1800 Btu/lb
Coal Heating Value	10,000 Btu/lb

Table 4-3. Cost and Fuel Savings, Electric Utilities Gas Steam Generator [56,57]

Unit Size (MW)	Capacity Factors (%)	Fuel Burned (Btu/hr) × 10^6	Capital to Modify ($) × 10^6	Break-Even Preps ($/MM BTU)	Local Preps ($/MM BTU)	Regional Preps ($/MM BTU)	Local Savings ($1YR) × 10^6	Regional Savings ($1YR) × 10^6	Gas Burned (cf/yr) × 10^9	Gas Saved (cf/yr) × 10^9	Oil Used (bbl/yr) × 10^6	Coal Used (ton/yr) × 10^3
50	30	150	4.2	—	0.26	0.24	—	—	1.314	1.314	0.134	25.7
	60	300		0.387	0.26	0.24	0.334	0.386	2.628	2.628	0.268	51.5
	90	450		0.536	0.26	0.24	1.088	1.167	3.942	3.942	0.403	77.2
75	30	225	6.3	—	0.26	0.24	—	—	1.971	1.971	0.201	38.6
	60	450		0.387	0.26	0.24	0.501	0.579	3.942	3.942	0.403	77.2
	90	675		0.536	0.26	0.24	1.632	1.750	5.913	5.913	0.604	115.9
100	30	300	6.4	—	0.26	0.24	—	—	2.628	2.628	0.268	61.5
	60	600		0.387	0.26	0.24	0.667	0.773	5.256	5.256	0.537	103.0
	90	900		0.536	0.26	0.24	2.176	2.334	7.884	7.884	0.805	154.5
200	30	600	13.2	0.132	0.24	0.24	—	—	5.256	5.256	0.537	103.0
	60	1,200		0.483	0.24	0.24	2.554	2.554	10.512	10.512	1.073	206.0
	90	1,800		0.600	0.24	0.24	5.676	5.676	15.768	15.768	1.610	309.1
300	30	900	16.8	0.238	0.22	0.24	0.142	—	7.884	7.884	0.805	154.5
	60	1,800		0.536	0.22	0.24	4.983	4.667	15.768	15.768	1.610	309.1
	90	2,700		0.635	0.22	0.24	9.815	9.343	23.652	23.652	2.415	463.6
400	30	1,200	20.0	0.302	0.22	0.24	0.862	0.652	10.512	10.512	1.073	206.0
	60	2,400		0.568	0.22	0.24	7.316	6.896	21.024	21.024	2.147	412.1
	90	3,600		0.657	0.22	0.24	13.781	13.151	31.536	31.536	3.220	618.2
500	30	1,600	22.0	0.366	0.22	0.24	1.918	1.656	13.140	13.140	1.342	257.6
	60	3,000		0.600	0.22	0.24	9.986	9.461	26.280	26.280	2.684	515.2
	90	4,500		0.678	0.22	0.24	18.054	17.266	39.420	39.420	4.026	772.8

600	30	1,800	23.4	0.419	0.22	0.24	3.138	2.822	15.768	15.768	1.610	309.1
	60	3,600		0.627	0.22	0.24	12.835	12.204	31.536	31.536	3.220	618.2
	90	5,400		0.696	0.22	0.24	22.517	21.571	47.304	47.304	4.831	927.3
700	30	2,100	24.5	0.462	0.22	0.24	4.512	4.084	18.396	18.396	1.879	360.6
	60	4,200		0.648	0.22	0.24	15.747	15.011	36.292	36.792	3.757	721.2
	90	6,300		0.710	0.22	0.24	27.042	25.938	55.188	55.188	5.636	1,081.9
800	30	2,400	24.8	0.604	0.22	0.24	5.971	5.550	21.024	21.024	2.147	412.1
	60	4,800		0.669	0.22	0.24	18.879	18.039	42.048	42.040	4.294	824.3
	90	7,700		0.724	0.22	0.24	31.788	30.527	63.072	63.072	6.441	1,236.5

Oil Storage	10 days
COM Storage	20 days
Cost per Man Hour	$20
Fixed Charge Rate	19%
Cost of Power	$.05/kWh

The estimate of cost and fuel savings based on these parameters for a utility gas steam generator is shown in Table 4-3. In this table the COM preparation costs have been converted to cost per unit of fuel using capital amortization and operation costs. If more recent values for gas, oil and coal prices are substituted, with their much higher differentials with respect to coal, it is clear that the investment value increases.

Kuehnel and Blake [59] at MITRE essentially concur in their similar analysis, which includes the additional considerations of tax savings as deferred interest charge and the stack emission control costs. Table 4-4 presents their cost assumptions and Table 4-5 their results on break-even oil-coal price differentials. These are the differences that would yield the required investment return for new utility investments. Some typical projections of oil and coal prices, which would give rise to the price differentials, are shown in Table 4-6. It is clear that current oil-coal price differentials indicate an attractive investment.

In both of the foregoing evaluations, investment value of the COM retrofit is based almost entirely on the oil-coal price differential. Moreover, that differential itself is presumed to grow faster than general inflation as reflected by the 4-percentage point difference in oil and coal escalation. This may tend to overevaluate the present value of the return from fuel savings more than, say, ten years in the future. A possibly less conventional analysis, which would reflect more accurately the most likely technical prospects, may be more valuable as an assessment of the COM development effort. This analysis would presuppose that eventually oil would be produced from coal at some proportionate premium over the coal price. Also, the COM itself would be prepared at a central facility serving a local region because of all the clear economies of scale in handling coal weighed against the difficulties that would be encountered at individual sites. For example, individual wintertime coal storage in cool climates will be subject frequently to huge expenditures in breaking up exposed coal piles frozen by spells of severe weather. A large facility will be able to avoid the problem with a more constant flow of the coal inventory and more closely controlled storage conditions.

We would then have to evaluate a slightly more complicated economy involving the COM preparation middleman. His profit must be realized from his price for the COM compared to the capital investment for the

Table 4-4. Assumptions of MITRE Economic Evaluation for COM [59]

COM Mixture Characteristics	Weight Percent	Energy Percent
Oil	44.0	55.2
Coal	50.0	44.6
Additive	.2	.2
Water	5.8	—

COM Cost Parameters—1977 Dollars

Plant Size (MWe)	100	200	400
Boiler Retrofit ($/kW)	35.0	28.5	23.2
Stack Emission Control ($/kW)	40.0	32.5	26.4
COM Preparation Plant ($/kW)	66.8	52.5	84.4
Total Investment ($/kW)	141.8	113.5	90.9
O & M Costs			
Manpower $/kW/yr	3.36	1.68	1.17
Utilities $/ton of COM	2.00	2.00	2.00

Financial Parameters

Remaining Boiler Life		20 Years
Tax Life of Preparation Plant		14 Years
Capital Structure and Required Return on Investment		
50% Debt	@	8% Return
50% Equity	@	13% Return
Inflation Rate		5%
Income Tax Rate		50%

Delivered Fuel Prices
Fuel Prices and Escalation
Price—1977 Dollars

Fuel	$/Unit	$/10⁶Btu	Escalation and Inflation
Coal—1.0%S	33.80–52.00	1.30–2.00	6.0%
Oil—1.0%S	14.14/bbl	2.30	6.9%
Oil—0.5%S	14.70/bbl	2.39	6.9%
Additive	4.70/gal	45.45	5.0%

plant. The utilities then have a somewhat smaller fuel savings balanced against a much smaller capital investment for the COM-burning retrofit. Some of the details involved in the consideration of an offsite 50% COM preparation plant have been discussed conceptually by Bergman et al. [52], who analyzed the case of small industrial plants. The COM, using extracted oil, would be prepared centrally. With a boiler capacity of only 100,000 lb/hr of steam, onsite COM preparation clearly would be unec-

Table 4-5. COM Break-Even Oil-Coal Price Differentials (¢/million Btu) [59]

Plant Size		Oil Price Escalation		
	Dollars/kW	Low	Base	High
400 MW				
Best estimate	90.90	$0.73	$0.50	$0.30
20% increase	109.08	$0.78	$0.55	$0.35
200 MW				
Best estimate	113.50	$0.83	$0.59	$0.39
20% increase	136.20	$0.89	$0.65	$0.45
100 MW				
Best estimate	141.8	$0.98	$0.74	$0.53
20% increase	170.16	$1.05	$0.82	$0.62
Cost Savings per $0.10 increase		$0.044	$0.04	$0.037

Table 4-6. MITRE Metrek COM Cost Model—Projected Fuel Prices ($/million Btu) [59]

Year	Coal	Oil LS	Oil VLS	Additive	COM[a]
1977	1.35	2.11	2.20	45.45	1.89
1978	1.42	2.26	2.35	47.72	2.01
1979	1.50	2.42	2.51	50.11	2.18
1980	1.59	2.58	2.68	52.61	2.27
1981	1.68	2.76	2.87	55.25	2.42
1982	1.77	2.94	3.05	58.01	2.56
1983	1.87	3.10	3.21	60.91	2.70
1984	1.98	3.26	3.38	63.95	2.85
1985	2.09	3.44	3.57	67.15	3.00
1986	2.21	3.63	3.76	70.51	3.17
1987	2.33	3.82	3.96	74.04	3.34
1988	2.46	4.03	4.17	77.74	3.52
1989	2.60	4.25	4.40	81.62	3.71
1990	2.75	4.47	4.64	85.71	3.92
1991	2.91	4.72	4.89	89.99	4.13
1992	3.07	4.97	5.15	94.49	4.36
1993	3.24	5.24	5.43	99.22	4.59
1994	3.43	5.52	5.72	104.18	4.85
1995	3.62	5.82	6.03	109.39	5.11
1996	3.82	6.14	6.35	114.86	5.39
1997	4.04	6.47	6.69	120.60	5.68
1998	4.27	6.82	7.06	126.63	5.99
1999	4.51	7.19	7.44	132.96	6.32

[a]COM price of fuel components only. All preparation costs excluded.
Oil LS contains 1.00% sulfur.
Oil VLS contains 0.5C% sulfur.
Coal contains 1.00% sulfur.

onomical. They propose an offsite central COM facility based on operations at the coal mine, as shown in Figure 4-1. For such a plant, operating at a capacity of 30×10^6 bbl/yr of COM, the required capital and operating costs are shown in Tables 4-7 and 4-8. These imply a price of \$1.88–\$1.94/million Btu heating value for the COM, using a No. 6 fuel oil price of \$2.16/million Btu. This cost is somewhat higher than the Eberle et al. [56,57] and Kuehnel and Blake [59] analyses and, therefore, may not represent the lowest-price supplier. Bergman et al. [55], estimated that then the retrofit for COM firing of a 100,000 lb/hr boiler would not clearly meet investment return requirements. However, continued extra escalation of the fuel oil price for a few years would create the circumstances for a profitable investment. For example, the June 1979 OPEC increase of up to 50% in the price of crude oil increased the gross fuel savings due to the use of COM at 50% coal by up to 25%. Since extra COM preparation operational costs imply an even higher proportionate increase in net fuel savings, this new circumstance may easily change a marginally unfavored investment to a clear favorite.

There is some evidence that this change was accomplished by the December 1979 of oil price rises, which moves the amount coal-oil comparison to the late 1980s on the projection of Table 4-6. Hickman [58]

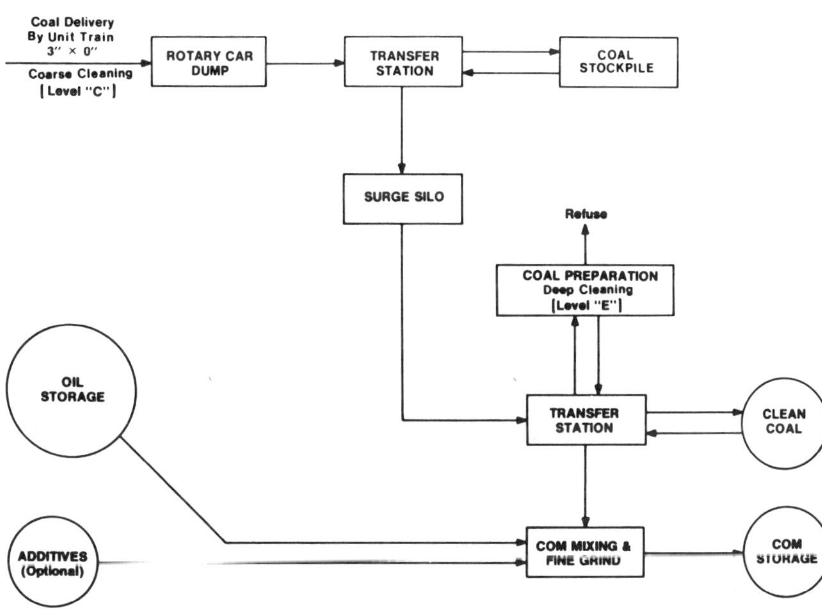

Figure 4-1. Offsite COM/coal preparation plant [55].

Table 4-7. Total Estimated Capital Requirements for 30 Million bbl/yr Offsite COM Plant [55]

Item	Cost
Coal Receiving, Storage, Reclaim	$ 3,206,200
Liquids Storage	35,172,000
COM Preparation	15,366,100
Plant Facilities	4,045,800
Plant Utilities	5,999,000
Total Construction	$63,789,100
Interest During Construction	7,654,700
Working Capital	
Facility and equipment	5,103,100
Oil inventory	11,076,300
Coal inventory	7,332,800
Additive inventory	900,000
Total Investment	$95,856,000

Table 4-8. Distribution of Annual Operating Costs of 30 Million bbl/yr COM Plant [55][a]

Cost Component	Percentage
Oil	69.3
Coal	23.0
Additive	3.2
Operation & Maintenance	2.3
Fixed Cost	1.7
Power	0.5
	100.0

[a]Annual operating cost of COM plant is $332,100,000.

has cited a sharp increased in Forney's industrial customers, who are interested in COM burners. Many of them plan to utilize the service offered by CoaLiquid (see Chapter 5) for provision of centrally prepared COM. Notably there is still no interest being expressed by utilities.

CHAPTER 5

REVIEW OF COAL-OIL MIXTURE PROGRAM STATUS

At present, the main developmental programs in coal-oil mixtures can be divided into the following categories:

- U.S. government-sponsored industrial development projects
- University research projects
- Other U.S. projects under private industrial sponsorship
- Foreign projects

U.S. government promotion of COM utilization clearly is moving from an era of direct support of development to a combination of mandatory fuel use laws and tax incentives. Notably it is anticipated that COM will be one of the more important "alternative fuels" in U.S. energy acts. The following section will describe the technical developments in the foregoing categories with respect to their current sponsorship levels. The most noteworthy features of a cross comparison of these categories is a plainly rising level of independent, privately sponsored developments and a startling rise in the extent and variety of Japanese efforts in the foreign sector.

Current status and principal thrusts of the various projects in all categories are summarized in the following subsections. An outline of the projects by organization and subject is provided in Table 5-1.

U.S. GOVERNMENT-SPONSORED INDUSTRIAL DEVELOPMENT PROJECTS

Most of the current U.S. government-sponsored projects are continuations of those already described in some detail in Chapters 2 and 3. Sponsorship has been terminated on the efforts of Accurex, Florida Power, and Gulf and Western. However, after completion of the Florida Power demonstration program on an operating boiler, the utility has continued

Table 5-1. Current COM Development Programs

(a) COM Boiler Plant Tests
Pittsburgh Energy Technology Center
Atlantic Research Corp.
New England Power & Service Co. (NEPSCO)
Tennessee Valley Authority (TVA)
Accurex Corp.
Appollo Technologies, Inc.
New Brunswick Electric Power Commission
Montreal Engineering
Mitsui Engineering & Shipbuilding
Babcock-Hitachi
(b) Specialized COM Burner Tests
Brookhaven National Laboratory
Wayne State University
Mitsubishi Heavy Industries
Japanese Electric Power Development Co.
(c) COM Stability
Germantown Laboratories
University of Massachusetts
University of Kentucky
(d) COM Pumping and Transport
Argonne National Laboratory
Resource Engineering, Inc.
Transamerica de Laval, Inc.
(e) COM Rheological Properties
Tufts University
Air Products/Wheelabrator-Frye
Adelphi University
(f) COM Preparation, Complete Plant
Methacoal Corporation
CoaLiquid Inc.
SWECO, Inc.
(g) COM Grinding & Mixing
Penn State University
Florida Power Corp.
Dravo Corp.
British Petroleum
University of Toronto
Kawasaki Heavy Industries
Hitachi Shipbuilding & Engineering
(h) COM Additives Development
Dow Chemical Corporation
Nippon Oil and Fat
Lion Fat and Oil
Kao Soap
Dai-ichi Kogyo-Seiyaku
(i) COM Economics
Applied Combustion Technology, Inc.

Table 5-1, continued

State of Florida
University of Central Florida
Davy McKee Corporation
MITRE Corp.
(j) COM Injection for Smelting
Interlake Corp.
Inland Steel Co.
Sumitomi Metal Industries

its partnership with Dravo to improve COM properties. The following projects continue to have some current DOE support as indicated:

1. Pittsburgh Energy Technology Center (in-house). A team of Bellas, Pan, Mathur and Bienstock is now conducting COM use tests on a 700-hp watertube boiler with coal contents up to 50%. Ekmann and Bienstock are continuing their studies of COM stability and the influence on oil chemistry, viscosity and the presence of water as additive.
2. Atlantic Research Corporation. Schefee and McHale are continuing to develop COM with high coal solids loadings, emphasizing a rising proportion of water. They are currently evaluating combustion characteristics of 65% and 70% water slurries of bituminous coal.
3. Germantown Laboratories. Cherry and Stokes are now experimenting with high-intensity mechanical mixing devices to enhance the stability of COM up to 45% coal loading with their K-scan unit used to evaluate the stability.
4. New England Power Service Company (NEPSCO). Utilizing partial DOE sponsorship, NEPSCO is pursuing several collaborative efforts. With Dooher's team at Adelphi, Tsai and Batra of NEPSCO are studying both the rheological and combustion properties of coal-oil-water emulsions. With Kurtzrock of the Pittsburgh Energy Technology Center, Batra of NEPSCO has established a committee to set standards for COM formulas and associated handling equipment. With 50% DOE sponsorship, Dunn of NEPSCO has started a feasibility test on an 80-MWe boiler using 30% COM.
5. Interlake Corporation. With 50% DOE sponsorship, a project on the use of COM injection to supply carbon to blast furnace melts has continued under Keyser. Most of the equipment for COM preparation has been completed and early operational tests are anticipated.
6. Argonne National Laboratories and Resource Engineering, Inc. This is a new DOE-sponsored project. Harkness and Petrovic have been studying COM pipeline characteristics and the possible advantages of coal-water or coal-methanol as a conveyance medium.

UNIVERSITY RESEARCH PROJECTS

The university work remains centered on small-scale laboratory work aimed at further elucidating the detailed physical properties of COM in the various processes it must undergo in use. Work at Northwestern on single-drop combustion and at the University of Texas on a lignite-alcohol slurry has been concluded. The following projects are under DOE sponsorship through the Pittsburgh Energy Technology Center:

1. Brookhaven National Laboratory. Butcher, Krishna and Pucci are studying the atomization coal slurry jets with both oil and water as particle dispersive media. They represent the process as a sequence of fluid sheets and columns breaking up into droplets.
2. Wayne State University. Carmi and Ghassemzadeh are studying COM atomization in experiments using elaborate photographic records of the process.
3. Pennsylvania State University. Luckie and Bagga are continuing a study of coal grinding according to kinetic parameters that have been defined to represent the process.
4. Tufts University. A team under Viola has been studying rheological properties of coal-oil mixtures, with emphasis on the important influence of the surface properties of coal over both aggregative and sedimentative stability.

OTHER U.S. PROJECTS

A large number of other projects are proceeding on all aspects of COM development. Only two of the nonindustrial projects are sponsored by DOE. The others are all privately supported and clearly aimed at specific near-term utilization of COM. The most noteworthy effort among the following is that of CoaLiquid in establishing a commercial-scale central COM plant.

1. Tennessee Valley Authority (TVA). Smith of TVA, aided by Cox of the University of Tennessee, and Berbridge et al. of Accurex, are testing the use of COM instead of oil in the light-off system (ignition) of a 2.6-GW coal-fired utility boiler.
2. Applied Combustion Technology. Varney and Tuliano of Applied Combustion, Woerner of the State of Florida and Worbs of the University of Central Florida are assessing the market potential for COM in a state supported study.

3. MITRE Corporation. Foo, Blake and Jamgochian are continuing studies, under DOE sponsorship, of the market potential for COM using comprehensive financial analysis of the entire conversion process. Jamgochian, Foo and Sabadell have developed a test plan for the NEPSCO COM combustion project.
4. Davy McKee Corporation. Under DOE sponsorship, G.A. Christie is examining the economic rationale for COM use, particularly as influenced by the most recent round of oil price increases.
5. Appollo Technologies. Mahoney and Marker are studying the effects of conversion to COM using a pilot-plant facility with capacity of 250,000 Btu/hr. This pilot facility has been instrumented much more extensively than in other such tests for such influences as chemical additives, temperature profiles and environmental effects.
6. Methacoal Corporation. A patented COM process named Carbonaceous Fuel-in-Oil Suspensoids (CFOS) has been developed by L. J. Keller and is now available for sale under various terms.
7. University of Massachusetts. Roswell and Marlow, with Ysai et al. of NEPSCO, in a NEPSCO-sponsored study, have studied sedimentation and subsidence of COM using the sedimentation column measurement.
8. Air Products/Wheelabrator-Frye. In a joint venture, Foster and Gupta, aided by Dooher of Adelphi and Kelly of Villanova, are studying rheological properties of COM that incorporate solvent-refined coal.
9. University of Kentucky. Coburn is testing the long-term stability of coal-oil-water emulsions in 55-35-10 proportion, which have been prepared by ultrasonic emulsification.
10. Dow Chemical Corporation. Meyer has been studying the use of water to stabilize COM against settling by acting as a flocculating third phase.
11. Florida Power Corporation and Dravo Corporation. These two companies have formed COMCO as a joint venture to continue work on a fluid mill grinding approach for preparing ultrafine (100%-325 mesh) COM. Cochran and Sell are the investigators.
12. Transamerica de Laval, Inc. Mabes is experimenting with the use of two- and three-screw positive displacement pumps to survey COM, concentrating on special points of wear and their influence on volumetric efficiency.
13. CoaLiquid, Inc. Under G.T. Hawkins, a central COM preparation plant, located in Shelbyville, Kentucky, has been available since June 1979. It is producing 2500 gph at startup and can be expanded to 5000-gph capacity.
14. Inland Steel Company. In a company-sponsored effort, Holowaty, Landsly and Wilson have been injecting a 50-50 coal-oil slurry as partial replacement for Bunker C oil in a blast furnace to assess its commercial prospects.

15. SWECO, Inc. A special effort, privately sponsored, is led by T.W. Summers, to prepare, in the most energy-efficient manner possible, COM slurries with solids content greater than 50%, anticipating that the near future energy costs will soon make it the most economical.

FOREIGN PROJECTS

At the recent MITRE-coordinated symposium on COM combustion, foreign interest was evidenced in the presentation of papers from the United Kingdom, Canada and Japan, and attendance by representatives from France, Sweden, Spain, etc. It was noticed that there were no representatives from West Germany. By far the largest number of contributed papers came from Japan. Those mentioned below are in fact consolidations, which were requested by PETC to avoid overwhelming dominance at the conference by Japanese work. They are described by the subject area under which the original submissions were consolidated.

United Kingdom

- British Petroleum
  Wall, Veal and Groszek have been grinding coal in fuel oil without additives to achieve high-temperature stability.

Canada

- New Brunswick Electric Power Commission and Montreal Engineering
  Davies et al. of New Brunswick and Wang and Natarajan of Montreal Engineering have been testing COM combustion on a 10-MW unit assessing the technical-economic prospects for COM use in Canadian utilities.
- University of Toronto
  Trass and Papachristodoulou have been grinding coal in fuel oil at 42% solids with a novel Szego mill.

Japan

- Mechanisms of COM Stabilization
  Nippon Oil and Fat, Lion Fat and Oil, and Kao Soap have been

using surfactants to stabilize COM. Laboratory tests on settling were checked with pilot-plant operations.

- Additives for COM Stabilization
  A very large number of additives (1300) were tested for stabilizing properties on several kinds of COM by Dai-ichi Koggo Seiyaku and NEOS.
- Fine COM Preparation
  The use of a variety of grinding techniques to confer stability on COM is being investigated by Kawasaki Heavy Industries and Hitachi Shipbuilding and Engineering. Pilot tests of COM with 50–55% coal by weight are being carried out by Babcock-Hitachi, Ishikawajima-Harima and Mitsubishi. Experiments involve pumping, piping, storage and tanker transport.
- Coarse COM System
  Pilot-plant tests on coarse COM are being carried out by Mitsui Engineering & Shipbuilding and Babcock-Hitachi.
- COM Combustion
  Atomizer performance and combustion of COM spray are being studied by Mitsubishi Heavy Industries under a team headed by Y. Fujima. The Japanese Electric Power Development Company is working with Forney International to test an air atomized burner on 50/50 COM, identical to those used by General Motors and the New England Power Company.
- COM Injection in Blast Furnace
  Research on COM injection on substitute for oil is underway at Sumitomo Metal Industries by a team under T. Miyazaki. The COM used is obtained at 50/50 from the Japanese Electric Power Development Company.

APPENDIX A

CHARACTERIZATION OF VISCOUS FLOW PROPERTIES

The linear Newtonian understanding of viscosity is illustrated in Figure A-1. Consider that in a fluid we can identify two parallel plates, each of area A, separated by a distance, h. A force, f, is applied to the upper plate perpendicular to this distance to impart and maintain a constant velocity, v. Then,

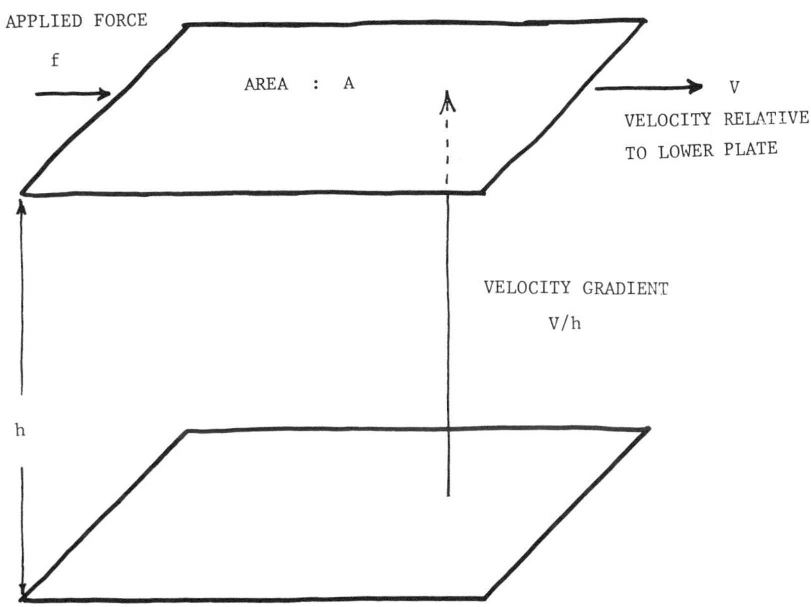

$$f = NA \cdot v/h$$

where N, VISCOSITY, DEFINED AS

$$N \equiv (f/A) \div (V/h)$$

Figure A-1. Newtonian viscosity.

in essence, we have an applied shear stress, f/a, between the plates resulting in a rate of shear strain equal to the velocity gradient, v/h. The proportionality between these two, which is measurable, is the Newtonian viscosity. Its dimension as grams per centimeter-second is known as the poise (p), after Poisseulle. In the figure, N is the symbol for viscosity relating applied force to rate of shear as

$$f = N\,A\,v/h$$

To arrive as some physical appreciation of the ordinary magnitude of the viscosity coefficient we can bear in mind the following representative values of these common fluids at 20°C.

Pentane	0.2395 cp
Water	1.009 cp
Glycerol	1.069 p
Light Lubricating Oils	0.025–1.5 p
Medium Lubricating Oils	1.5–3.5 p
Heavy Lubricating Oils	3.5–20 p
Castor Oil	15.14 p

It was apparent with the earliest work on COM that under operating conditions its fluidity would be comparable to that of the heavy lubricating oils. Generally the flow is in the laminar range where the classical Hagen-Poisseulle relation would govern its flow through a pipe [60]. For a round pipe of length, l, and diameter d, with applied pressure differential, P_d, a fluid of viscosity, N, would flow through at the volumetric rate

$$Q = \pi d^4\, P_d \div (128\ Nl)$$

Hence, required flowrates can easily be met with appropriate diameter. For a given facility, however, it would vary inversely with viscosity.

In most current U.S. work viscometers are commonly read as the ratio of viscosity to the fluid density in the dimension centimeter squared per second, denoted as the stoke, or kinematic viscosity. As with the poise, the most usual unit is centistoke. Since most fluid densities are around 1 g/cm^3, kinematic viscosity in centistokes will be approximately equal to viscosity in centipoises. The densities for coal-oil mixtures increase with concentration but do not climb above 1 by any substantial amount.

Gases, simple organic and inorganic liquids, true solutions, dilute suspension and emulsions are generally Newtonian in character, with shear strain rate directly proportional to applied shear stress. The fluids that deviate from this direct proportionality, that is, of non-Newtonian viscosity, include concentrated colloidal, other particulate systems in fluid media, and solutions of

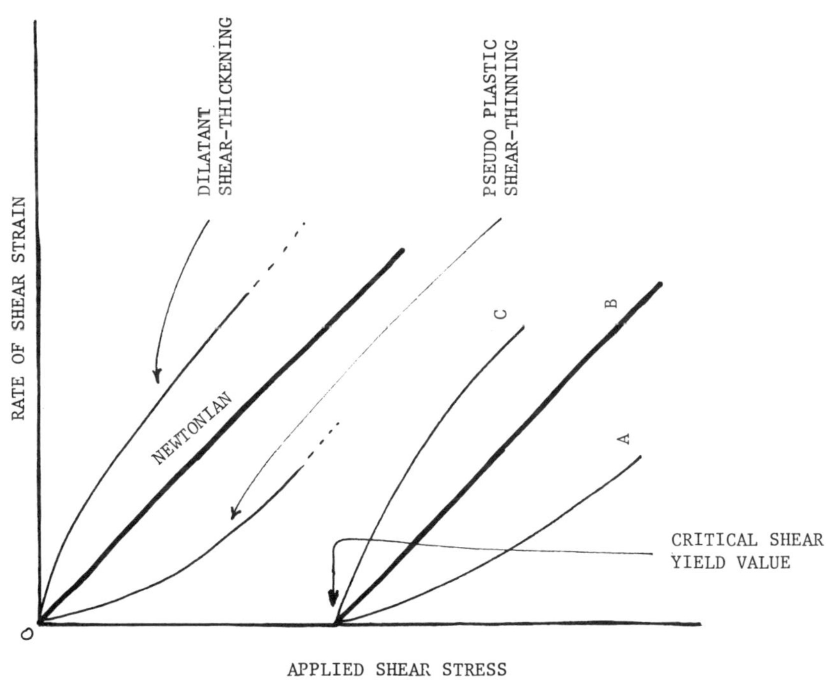

Figure A-2. Non-Newtonian viscous characteristics.

polymers and other macromolecoles in dispersed systems. Curves representing types of non-Newtonian behavior are shown in Figure A-2. When the shear strain rate increases proportionately more with added stress increments, the fluid is called pseudoplastic or shear-thinning. By contrast, for proportionately lower shear strain increase with added stress increments, the fluid is called dilatant or shear-thickening.

There are also materials that will not flow until a critical shear stress is exceeded, after which they may exhibit the same classes of behavior as represented by the current A, B, C of the figure. In this category, materials of type B are called plastic or a Bingham body, A and C being considered pseudoplastic and dilatant, respectively. Finally there are fluids whose viscous properties change with time after the onset of applied shear stress. If they become more fluid with increasing time they are called thixotropic. The opposite effect is called rheopexy. Such time-dependent effects may be spontaneous, reversible or irreversible. Typically, the apparent viscosity of a thixotropic fluid decreases logarithmically with time toward a limiting value.

APPENDIX B

SUMMARY OF FINAL REPORT ON GENERAL MOTORS COM PROJECT, 1975-1977

(EXTRACTED FROM GENERAL MOTORS FINAL REPORT, OCTOBER 1977)

CONCLUSIONS AND RECOMMENDATIONS

The overall result of this Powdered Coal-in-Oil Mixtures Program is that a sufficiently stable coal-oil mixture can be prepared continuously and successfully fired up to 75% of maximum boiler rating of an existing (120,000 pph) gas-designed industrial boiler. Many problems were encountered in the slurry preparation plant and most were solved. The remaining problems are not considered to be inherent to the use of coal-oil fuels. It is believed that with some preliminary small-scale testing these problems can be eliminated by proper design of future preparation plants.

From an overall point of view, the air-atomized burners and the boiler operated well on the coal-oil mixture. The coal carbon burnout was about 98%. There was no excessive buildup of ash or slagging on the boiler tubes or on the furnace floor and no evidence of erosion of boiler tubes. Measurements of stack emissions showed that these could be treated by standard technology. There was no major deterioration of boiler efficiency.

From a more detailed viewpoint, the boiler steam output was unstable and sometimes did not follow the indicated fuel consumption rate. The boiler tripped off frequently. This behavior was traced to fuel pressure. The cause of the pressure variations could not be determined completely, but variations in viscosity of the coal-oil mixture and transient flow constrictions are the most likely candidates.

Based on the data produced in this study and including installations of the necessary emission control devices, an economic analysis showed the coal-oil mixture concept marginally attractive, depending on current local coal and oil costs.

These conclusions will now be reiterated point by point with more detail and with reference to the appropriate sections of the main report for supportive data. Finally, the recommendations for further work will be presented.

CONCLUSIONS AND OBSERVATIONS

General

1. The demonstration was carried out in two phases: In Phase I, from March 1976 through June 1976, over 250 hours of operation were accumulated with the combustion of 60,000 gallons of a nominal 35 wt % coal, coal-oil mixture. In Phase II, from March 1977 through April 1977, 494 hours of operation were accumulated with combustion of over 250,000 gallons of a nominal 50 wt % coal, coal-oil mixture.

2. In both phases of demonstration, startup problems and delays in delivery of equipment and/or coal postponed operations until the steam load on the plant had decreased so that full steam load on the No. 5 boiler could not be obtained. The maximum sustained steam load of about 90,000 pph of the maximum rating of 120,000 pph was obtained in both Phase I and Phase II operations. A steam load of greater than 60,000 pph was maintained over 32% of the operation in Phase II.

Coal-Oil Mixture Characteristics

1. There are several methods (e.g., chemical, mechanical) by which coal-oil mixtures of varying degrees of stability can be made.

2. The coal-oil mixture, including any additive, must be tailored for both the coal and oil used.

3. Stable coal-oil emulsions can be prepared while using oils having a wide range in viscosities, coals with nominal particle sizes of 15–75 μ and coal weight content up to 50%.

4. The No. 6 fuel oils tested showed a very wide range of viscosities at low temperatures, but at 160°F the viscosities of all the oils were about the same.

5. Slurry stability is very sensitive to the particle size distribution of the coal.

6. Slurry stability tends to increase with coal concentration.

7. Increasing temperature greater than 150°F tends to promote settling of the coal.

8. Mechanical means of preparation of the coal-oil mixture includes intensive mixing as well as fine grinding of the coal.

9. Standard equations for Newtonian fluids (for Headstrom numbers < 1000) can be used for design purposes for coal-oil mixtures. However, the viscosity of the mixtures is very dependent on coal concentration. Pressure drop in pipes scales with diameter as D^4, but it is much higher for coal-oil mixtures than for No. 6 fuel oil.

10. Several of the coal-oil mixtures studied can be stored for periods of weeks with little buildup of hard, nonpumpable material in the bottom of the tank.

11. For usual pipe velocities (< 10 ft/sec), erosion will be minimal in

straight pipe, standard fittings, globe and ball valves. However, fittings with sharp changes in stream lines, like partially closed gate valves and relief valves, did show appreciable erosions.

12. Tests showed only a slight difference in corrosion rates between coal-oil mixtures and No. 6 fuel oil.

13. Most of the coal-oil mixture properties (e.g., ash and sulfur content, heating value) are the weighted average of the individual properties of the coal and oil. Surprisingly, however, the data show that the specific volumes are not additive.

14. For the coal and oil selected for Phase II demonstration, a test for coal-oil mixture stability at 150°F showed an optimum at the composition 6.5 wt % water, 0.2 % additive, 50.0 wt % coal and 43.3 wt % No. 6 fuel oil.

Mixing (Phase II Demonstration)

1. The coal-oil mixture selected for the Phase II demonstration could be made by first emulsifying the oil, water and additive and then mixing this emulsion with the pulverized coal using normal mixing procedures.

2. Because of this simplicity of slurry preparation it was possible to build and successfully operate a continuous slurry preparation plant at the Saginaw test site.

3. Improvements in control of slurry preparation are needed as the final product to the boiler varied by as much as ±10 wt % coal.

COM Burning Characteristics

1. Coal-oil mixtures burn in a manner very similar to No. 6 fuel oil.

2. The coal-oil mixture demonstration tests verified many of the observations made in the preliminary combustion tests on the small furnace: same fuel:air ratio, stable flames, high emissivity, same opacity, no ash buildup, no slagging of the ash, trouble-free operation on the air-atomizing burner and some coking on the steam atomizing burner.

3. Throughout the program, beginning with the preliminary combustion tests, the fire appeared very similar to that of No. 6 fuel oil except for its dusty appearance. There was no evidence of a coal burnout plume.

4. The air-atomized burner performed very satisfactorily in the range of 35 to 50 wt % coal COM.

5. The steam-atomized burner performed reasonably well as coal concentration up to 40 wt % coal but tended to plug and produce unstable flames from 40–50 wt % coal.

Boiler Performance

1. Inspection of the furnace of each demonstration showed no evidence of damage and erosion to the boiler surfaces.

2. Ash buildup on the floor was minimal (1 inch), and the largest deposits were in the corner. The appearance of the ash deposits was similar to that of drifted snow in that the ash accumulated behind any obstacle to free flow or in any dead volume until only streamline flow was established. The ash deposits may very well have represented a steady-state accumulation, with zero deposition rate over the last parts of the tests.

3. The special soot-blower arrangement developed for ash removal from the furnace floor was not needed.

4. There was no ash in the convection passes, and the ash analyses revealed no residual carbon.

5. The stack breeching and Lundstrom preheater showed no signs of excessive wear or ash deposition.

6. The air-atomized burners had a crust of material on the gun bodies and on the lower portion of the diffuser.

Coal Concentration Devices

1. The ultrasonic and dielectric devices developed to continually measure the coal concentration of the coal-oil mixture to the boiler both worked satisfactorily as prototype instruments.

2. Continuous measurement during the permitted 4½ hours of operation showed three types of variations in the slurry coal concentration: (1) variations of ±0.25 wt % with a duration of about 30 seconds; (2) variations of ±2 wt % with a duration of order 10 minutes; and, finally (3) a long-term drift of order 5 wt % in coal concentration over a period of about 2½ hours.

3. The third variation above resulted in a flameout of the boiler and strongly suggested that part of the boiler instability is caused by greatly increased viscosity (and reduced flows) with increased coal concentrations.

4. Even though a period of online demonstration of the devices was very short, it was concluded that their use can be very valuable in the control and diagnosis of boiler operation on coal-oil mixtures.

RECOMMENDATIONS

Preface

The following recommendations are in two categories. The first recommends studies on a smaller scale to provide additional detailed information needed for future demonstration-scale tests. Based on the above conclusions, the results sought would avoid some of the above problems and, therefore, yield firmer data and conclusions. The second category comprises technical recommendations for demonstration-scale tests.

Of major importance is the need to establish good control of the fuel feed variables before extensive boiler testing is initiated. This is not achieved easily since the quality of control (i.e., allowable ranges of variation of these variables) must itself be established by preliminary boiler tests on the fuel feed. Again, to emphasize this point, the preliminary tests as well as the boiler rating and efficiency tests require periods of operation (hours) during which temperature, concentrations of coal and water, and supply pressure to the fuel feed are steady.

Instrumentation for boiler tests, particularly those for fuel flow measurement and control, needs careful study for use with coal-oil mixtures to determine the sources of error and malfunction. Additional instrumentation to continuously monitor all fuel feed properties is highly desirable.

Detailed boiler tests for efficiency and rating should be made with a fuel supply system that is sufficiently controlled and can ensure normal, steady steam flow over intervals of hours. Tests should be included that assess the effect of different boiler configurations and different coals and boiler surfaces.

Recommendations for Small-Scale Testing

1. Further testing should be carried out on various types of coal-oil mixtures to provide a better understanding of the mechanisms that enhance stability. Experiments to improve the stability of coal-oil mixtures should be continued. Tests on the more promising of the coal-oil mixtures should be conducted through the pilot-plant stage to better establish the scaleup factors.

2. Carefully check out the physical properties of the selected coal-oil mixture, not only at the design conditions and concentrations, but also over possible ranges of variance of these variables.

3. If possible, check out the burning characteristics of the selected coal-oil mixture on small-scale equipment.

4. Laboratory-scale coal-oil mixture preparation test facility should be built to further develop the technique of continuously making stable coal-oil mixtures of closely controlled composition and temperature and to help establish scaleup factors. It should be sized to use small-scale coal grinding and storage equipment in the range of 10 pph and 1000 lbm, with corresponding oil storage capability. Instruments or continuous recording of mixture properties should be incorporated as they are developed. New methods of determining mixture stability and composition should be applied as they become available.

5. A small-scale coal-oil mixture flow components and instrumentation test facility (CITF) should be built. A maximum flow capacity of about 10 gpm is sufficient to provide information useful for industrial-scale systems while retaining flexibility, low cost and care of disposal of test fluids. Determination of the causes of pressure variation, instrument malfunction and component wear should be a primary objective. Instrumentation should be provided to monitor for transient plugging. The test system should be used to develop improved instruments for continuous measurement of coal concentration,

viscosity and flow of COM. Several different storage tank configurations with capacities of about 1000 gallons should be incorporated. These should be used to determine the best methods of filling, heating, mixing and removal. Such studies should produce methods and designs to ensure a fuel flow of uniform properties at a specified value of pressure and temperature.

Recommendations for Demonstration-Scale Testing

1. At the earliest opportunity after site selection, conduct briefing seminars to both educate and motivate personnel at the site of the objectives, plans and importance of the project.

2. Future tests should be performed on coal-oil mixtures with different types of coal to study the effect on boiler operation (and on slurry stability) of lower ash fusion temperature coals.

3. The effect of different boiler configurations on the combustion characteristics of coal-oil mixtures should be tested.

4. The slurry preparation plant should be provided with two separate methods for determining each of the feedrates, the coal-oil mixture production rate and the boiler feedrate.

5. Coal grinding operations should be made a part of the total slurry preparation plant to ensure availability and adequate control of product quality. However, storage of pulverized coal should be minimized. Use of outside vendors for coal-grinding operations is adequate for small pilot operations.

6. The coal-oil mixture should be stored at a temperature below 150°F. The mixture should then be heated to a temperature of 160–180°F for transport to the boiler. The fraction of the hot boiler feed that is returned to the storage tank should be minimal.

7. Special care should be taken in the design and selection of equipment for throttling and safety relief of coal-oil mixtures. Use variable-speed, constant-displacement pumps. Use relief valves for safety only and then install with a warning that a valve has been used. Also, install the relief valve with sufficient additional valving and pressure gauges to check that the relief valve has reseated.

8. Pumps, piping and process equipment should be oversized to reduce pressure drop slurry flow and also allow a volume reserve in the system. Use ball or globe valves for throttling coal-oil mixtures.

9. Checkout should be made on any existing boiler control equipment that is expected to operate directly on the coal-oil mixture.

10. Coal-oil mixture storage tanks should be designed so that any material settled will be picked up immediately in the pump suction and either be sent to the boiler or recycles to the top of the storage tank. Capability for heating of the suction line to pumps should be provided.

11. Heating of the coal-oil mixture storage tank should be applied evenly and over as large an area as possible.

12. Means for agitation of the storage tank should be provided as a precautionary measure.

13. In future full-scale testing, signals from existing boiler diagnostic instruments should be connected to a multichannel strip-chart recorder. This would greatly aid and improve the subsequent analysis of the data. Also, the feasibility of continuously monitoring and including such other variables as O_2 and CO_2 concentration, and fuel pressure, temperature and coal content should be assessed in each case. The recorder should be continuously attended by test personnel who should note *all* significant conditions and events on the chart.

14. Before detailed efficiency tests are made in future coal-oil mixture demonstrations, sufficient preliminary testing should be done to establish reliable stable boiler operation over periods of at least three hours on industrial boilers.

Preliminary testing on the slurry preparation plant should be done to ensure a product of specified and constant composition, temperature and pressure to the boiler.

Following, or in conjunction with, the slurry plant tests, controlled changes in the composition, temperature and pressure of the slurry to the boiler should be made to evaluate the effect of each of these variables on the boiler operation.

Finally, when confidence in stable boiler operation has been established, then detailed efficiency tests should be performed.

15. A thorough test plan is essential for the success of a demonstration-scale testing program.

REFERENCES

1. Smith, H. R., and H. M. Munsell, "Improvement in Liquid Fuels," U.S. Patent 219, 181, (February 24, 1879).
2. Schroeder, W. C. "Mixtures of Oil and Coal in Boiler Furnaces," *Mech. Eng.* 64, (11): 793, (1942).
3. Demeter, J. J., C. R. McCann, A. T. Bellas, J. M. Ekmann, and D. Bienstock, "Combustion of Coal-Oil Slurry in a 100-HP Firetube Boiler," *Combustion* 49, (10): 31, (1978).
4. Manning, A. B., and R. A. A. Taylor, "Colloidal Fuel," *J. Inst. Fuel* 9: 303, (1936).
5. Manning, A. B., and R. A. A. Taylor, "Colloidal Fuel," *Trans. Inst. Chem Eng.,* 14: 45, (1936).
6. Adams, R. A., F. C. V. Holmes, and A. W. Perrin, "Colloidal Fuel Using Cracked Oil and High Carbon Residue," U.K. Patent 396, 432 (August 2, 1933).

8. Barkeley, J. F., A. B. Hersberger, and L. R. Burdick, "Laboratory and *on COM Combustion,* St. Petersburg Beach, FL, May 7-9 (1944), p. 21.

7. Nakabayashi, Y. "Outline of COM R&D in Japan," *Proc. 1st Int. Symp.* Field Tests on Coal-in-Oil Fuels," *Trans. Am. Soc. Mech. Eng.* 66, 185 (1978).

9. Jonnard, A. "Colloidal Fuel Development for Industrial Use," *Kansas State College Bull.* 30: 2 (January 15, 1946).
10. Barnard, W. N., F. O. Ellenwood and C. F. Hirshfeld. *Heat Power Engineering* (New York: John Wiley & Sons, Inc., 1935).
11. Flagg, H. "Colloidal Fuel at Armco, Middletown, Ohio," Proceedings of AIME National Open Hearth Conference (1944), pp. 47-58; (1945), pp. 306-309.
12. Rudzki, E. M., B. K. Pease, and T. H. Weidner, "Use of Coal-in-Oil Mixtures to Improve Open-Hearth Furnace Performance," *J. Inst. Fuel,* 154, (April 1965).
13. Decker, A. M. "Thermal Aspects of Blast Furnace Fuel Injection," *J. Metals,* 13, (1): 41, (1961).
14. Dietz, J. R. "Solid Fuel Injection at Weirton," *J. Metals* 15 (7): 499 (1963).
15. Stormont, D. H. "Fuel Oil-Coal Slurry: Cheap New Fuel for Making Steel," *Oil Gas J.* 60, (15): 78 (1962).
16. Holowaty, M. O., J. S. Kapitan, and N. A. Robins, "Toward Maximizing Fuel Oil Injection in the Iron Blast Furnace," *Ind. Heating* 40: 16 (November 1973).

17. Walsh, J. P. "Fuel Injection Blast Furnaces-Sharon Steel Corporation," *Blast Furnace Steel Plant* 53, (2): 143 (1965).
18. Freedman, S. I., C. B. Foster, and E. M. Jamgochian, "Technical Aspects of Coal-Oil Mixture Combustion," U.S. Dept. of Energy (CONF-780206 KC-90), February (1978).
19. Keyser, N. H., and R. P. Winters, "Interlake, Inc. Demonstration Project," *Proc. 1st Int. Symp. on COM Combustion,* St. Petersburg Beach, FL, May 7-9 (1978), p. 66.
20. Brown, A., Jr., Ed. "Final Report of the General Motors Corporation Powdered Coal-Oil Mixtures (COM) Program," G.M. Corp., Warren, MI (1977).
21. American Boiler Manufacturers' Association. "Analysis of Sales Data" (1976).
22. Foo, O. K. Personal Communication (1979).
23. Brown, A., Jr., "The General Motors Powdered Coal-Oil Mixture Program," *Proc. Coal-Oil Mixture Combustion Technol. Exchange Workshop,* Washington, DC, (1976), p. 125.
24. Hatano, M. "EPDC Com R&D Laboratory Tests and Pilot Plants," *Proc. 1st Int. Symp. on COM Combustion,* St. Petersburg Beach, FL, May 7-9 (1978), p. 113.
25. Dooher, J. P., R. W. Genberg, S. Moon, D. Wright, S. E. Jakatt, B. J. Gilmartin, Skura, Jr., and J. Lepore, "The Feasibility of Using Coal/Water/Oil as a Clear Liquid Fuel," *Proc. 1st Int. Symp. on COM Combustion,* St. Petersburg Beach, FL, May 7-9 (1978), p. 169.
26. Vanston, H. "METHACOAL—Waterless Coal Slurry," *Mech. Eng.* (April 1978), p. 47.
27. Luckie, P. T., G. N. Fletcher, T. K. Lau, and P. S. Bagga, "On the Grinding of Coal in Oil," *Proc. 1st Int. Symp. on COM Combustion,* St. Petersburg Beach, FL, May 7-9 (1978), p. 197.
28. Austin, L. G. "A Review Introduction to the Mathematical Description of Grinding as a Rate Process," *Powder Technol.* 5 (1971/72).
29. Sheppard, S. E. *J. in Eng. Chem.*, 13: 37 (1921).
30. Charm, S., and W. McComis, *Food Technol.* 19: 948 (1965).
31. Scheffee, R. S. "Development and Evaluation of Highly-Loaded Coal Slurries," *Proc. 1st Int. Symp. on COM Combustion,* St. Petersburg Beach, FL, May 7-9 (1978) p. 222.
32. Ekmann, J. M., and D. Bienstock, "Stability of Coal-Oil Mixtures," *Proc. 1st Int. Symp. on COM Combustion,* St. Petersburg Beach, FL, May 7-9 (1978), p. 273.
33. Sawyer, E. W. "The Use of Attapulgite Clay-Surfactant Gels to Stabilize Coal-Oil Mixtures," *Proc. 1st Int. Symp. on COM Combustion,* St. Petersburg Beach, FL, May 7-9 (1978), p. 317.
34. Sabadell, A., "Preliminary Evaluation of Commercial Continuous Instrumentation for the Measurement of Viscosity and Coal Concentration of Coal-Oil Mixtures," U.S. DOE, HCP/T2463-03, (October 1978).
35. Leffert, C. B. "Properties and Behavior of Some Coal-Oil Mixtures," *Proc. 1st Int. Symp. on COM Combustion,* St. Petersburg Beach, FL, May 7-9 (1978), p. 322.

36. Berkowitz, N., C. Moreland, and G. F. Round, "Pipeline Flow of Coal-in Oil Suspensions" (June 1963), p. 116.
37. Brown, R. A. S. "Pipelining Coal-in-Oil Slurries," *Coal-in-Oil Seminar Canadian Combustion Research Lab,* (1972), p. 19.
38. Wasp, E. J., J. P. Kenny, and R. L. Gandhi, "Solid-Liquid Flow Slurry Pipeland Transportation," *Trans-Tech Publ.* (1977).
39. Geraci, E. E. "Measurement and Control of the Coal in Coal-Oil Mixture," *Proc. 1st Int. Symp. on COM Combustion,* St. Petersburg Beach, FL, May 7-9 (1978), p. 260.
40. Kugel, R. W. "The Determination of Coal-Oil Mixture Stability," *Proc. 1st Int. Symp. on COM Combustion,* St. Petersburg Beach, FL, May 7-9 (1978), p. 300.
41. Cherry, N. H., and C. S. Stokes, "Use of Dielectric Properties in the Determination of COM Stability," *Proc. 1st Int. Symp. on COM Combustion,* St. Petersburg Beach, FL, May 7-9 (1978), p. 233.
42. Rowell, R. L., S. R. Vasconcellos, and J. R. Ford, "Investigation and Measurement of Stability in Coal-Oil Mixtures," *Proc. 1st Int. Symp. on COM Combustion,* St. Petersburg Beach, FL, May 7-9 (1978), p. 288.
43. Haake, Inc. "Haake Viscometers," *Haake* Catalogue RV-200/578/778 (1978).
44. Dunn, R. M. "The New England Power Service Co. Demonstration Project," *Proc. 1st Int. Symp. on COM Combustion,* St. Petersburg Beach, FL, May 7-9 (1978), p. 35.
45. Batra, S. K., and S. Y. Tsai, "Evaluation of Mixing Equipment for NEPSCO Coal-Oil Mixture Demonstration, *Proc. 1st Int. Symp. on COM Combustion,* St. Petersburg Beach, FL, May 7-9 (1978), p. 207.
46. Busch, C. F., and R. A. Brown, "Parametric Study of Coal-Oil Mixture Combustion," *Proc. 1st Int. Symp. on COM Combustion,* St. Petersburg Beach, FL, May 7-9 (1978), p. 183.
47. Schimizu, A. B., and T. C. Derbidge, "Demonstration of COM combustion in an Industrial Scale Boiler, *Proc. 1st Int. Symp. on COM Combustion,* St. Petersburg Beach, FL, May 7-9 (1978), p. 80.
48. Law, C. K., C. H. Lee, and N. Srinivasan, "Disruptive Combustion of Coal-Oil Mixtures Droplets," *Proc. 1st Int. Symp. on COM Combustion,* St. Petersburg Beach, FL, May 7-9 (1978), p. 90.
49. Beer, J. M. *Combustion Aerodynamics* (New York: John Wiley & Sons, Inc., 1972).
50. Rodriguez, L., and F. R. Sell, "Florida Power Corp/Dow Corp. Coal/Oil Composite Fuel Program," *Proc. 1st Int. Symp. on COM Combustion,* St. Petersburg Beach, FL, May 7-9 (1978), p. 57.
51. Essenhigh, R. H., C. Briceland, E. Goodzait, A. Kokkinos, T. Masuda, and H. E. Shell, "Thermal Efficiency of a Hot Wall Furnace Fired with Coal-Oil Mixtures at Low Concentrations (0-16%)," *Proc. 1st Int. Symp. on COM Combustion,* St. Petersburg Beach, FL, May 7-9 (1978), p. 153.
52. Ekmann, J. M., C. R. McCann, M. P. Mather, and D. Bienstock, "Parameter Studies of COM Combustion," *Proc. 1st Int. Symp. on COM Combustion,* St. Petersburg Beach, FL, May 7-9 (1978), p. 101.

53. Ehrlich, S. "EPRI's Interest in Coal Slurry Fuels," *Proc. 1st Int. Symp. on COM Combustion,* St. Petersburg Beach, FL, May 7-9 (1978), p. 270.
54. "Slurries; Coal/Oil," *Power,* S-22 (February 1979).
55. Bergman, P. D., J. I. Joubert, and D. Bienstock, "Economic Considerations for Industrial Firing of Coal-Oil Mixtures," *Proc. 1st Int. Symp. on COM Combustion,* St. Petersburg Beach, FL, May 7-9 (1978), p. 351.
56. Eberle, J. W., and R. H. Hickman, "Can Coal-Oil Mixtures Make it as Industrial Fuels," *Mech. Eng.* 100 (3): 25 (1978).
57. Eberle, J. W., R. H. Hickman, and F. M. Talmund, "Utilization of Coal-Oil Mixtures," *Proc. 1st Int. Symp. on COM Combustion,* St. Petersburg Beach, FL, May 7-9 (1978), p. 364.
58. Hickman, R. H. Personal Communication (1980).
59. Kuehnel, R. C., and J. C. Blake, "METREK Model for Financial Evaluation of Boiler Conversion to Coal-Oil Mixtures," *MITRE Corp.,* U.S. DOE HCP/T 2453-02 UC-90R (April 1978).
60. Dodge, R. A., and M. J. Thompson, "Fluid Mechanics" (New York: Book Co., 1937), *McGraw-Hill.*

INDEX